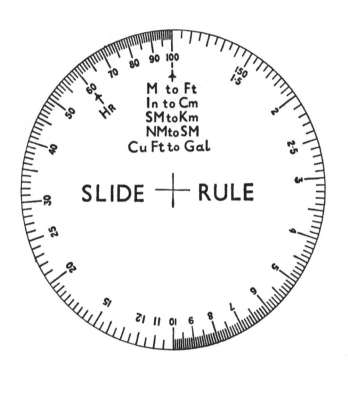

# MODERN MATHEMATICS

BY

### S. A. WALLING
*Senior Master R.N. (Ret.)*

AND

### J. C. HILL, B.A. (CANTAB.)
*Education Department,
Cambridge University Press*

CAMBRIDGE
AT THE UNIVERSITY PRESS
1948

## CAMBRIDGE
### UNIVERSITY PRESS

University Printing House, Cambridge CB2 8BS, United Kingdom

Cambridge University Press is part of the University of Cambridge.

It furthers the University's mission by disseminating knowledge in the pursuit of education, learning and research at the highest international levels of excellence.

www.cambridge.org
Information on this title: www.cambridge.org/9781316612668

© Cambridge University Press 1948

First published 1948
First paperback edition 2016

*A catalogue record for this publication is available from the British Library*

ISBN 978-1-316-61266-8 Paperback

# CONTENTS

[ iii ]                                                             I-2

# CONTENTS

[ iv ]

# PREFACE

In this book the authors assume that the student is likely to undertake more individual effort rather than to work under group control. The problems have been separated into different vocational or occupational types. In this way it is hoped to maintain interest and to encourage further progress in any particular sphere and to indicate the type of problem, found in industry or at work, which can easily be answered by elementary mathematics. There has not, in any instance, been an attempt to invent a problem to fit any particular fundamental rule; where a problem obviously suggests itself then it is included, otherwise it is omitted.

It is also assumed that ability to work correctly simple mechanical processes in fundamentals has been attained, so that mechanical exercises, as such, are reduced to a minimum, and there is no apology for the omission of such 'museum pieces' as 4 miles 3 furlongs 2 yards 2 feet 5 inches × 29.

Knowledge of simple tables of weights and measures is also assumed, although, for the most part, the employment of these various units is confined to the sphere to which, by usage, they belong. For instance, cement is bought and used by the ton or cwt. or fractions of a cwt. Quarters, stones or pounds are not used when dealing with cement.

Those students, therefore, who propose to earn their living in industry or trade should find in the various types of example something which should appeal and consequently instruct by reason of its utility.

In this respect the sections on Logarithms and Formulae are particularly important. The principles involved have been explained as clearly and simply as possible, and in the case of formulae brief descriptive text is introduced wherever these are used in order to stimulate interest and to illustrate a little of the wide application of the use of formulae in industrial and engineering work.

For simplicity of selection, the problems in the exercises are classified according to their application, thus:

A denotes that the problem deals with agriculture, farming, butchering, etc.

B denotes a problem in building, surveying or general constructional work.

Tr denotes a problem in transport (road, rail, sea or air).

T denotes a problem with technical application, including mechanical, electrical and structural engineering.

G denotes a problem of general interest.

The Tables of Logarithms and Trigonometrical Ratios printed on pp. 148–153 are included by the kind permission of the Cambridge Local Examinations Syndicate.

S. A. W.
J. C. H.

*May* 1946

# CHAPTER 1

# THE FOUR RULES

## Revision of Addition, Subtraction, Multiplication and Division of Numbers and Quantities

A quantity denotes a number of some definite unit or object; 10 tons, 15 seconds, 12 shillings are quantities.

Every calculation that has to be made to obtain the answer to a problem involves one or more of the above four rules.

Accuracy, therefore, is at all times essential. Always check an answer and wherever possible do this by an alternative method.

For example, when working an addition sum, do the addition, when checking, in a different order, so that any error made during the first calculation is less likely to be made in the second.

In multiplication, change the order of multiplying, when checking, for the same reason.

Extra care must be taken with problems that call for the use of large numbers, such as statistics of crops and acreage, census of population, etc.

## EXERCISE 1

Where necessary, reference should be made to the road map on p. 2 (fig. 1).

**1** (A). The wheat harvest in Great Britain, during 1938, amounted to 1,959,000 tons. In 1939 the yield was 317,000 tons less. What was the weight of the wheat crop in 1939?

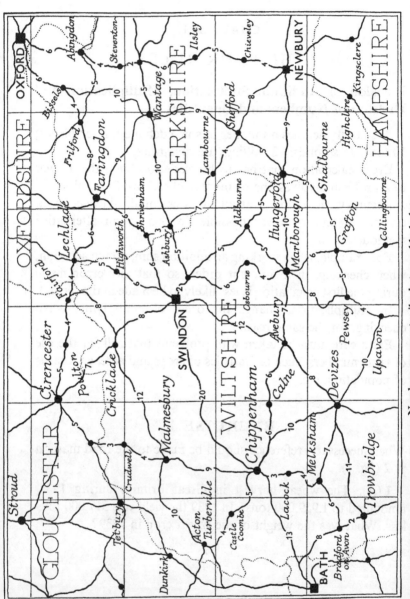

Numbers represent miles between black dots

**2** (Tr). You are in charge of a motor van and have to drive from Oxford to Bath. What would be your road mileage if you travelled (*a*) via Cirencester, (*b*) via Swindon, (*c*) via Hungerford?

**3** (T). If the internal bore of a cast-iron main sewer is 15 in., and the thickness of metal is 1 in., what is the external diameter of the sewer?

**4** (Tr). A vessel left Plymouth bound for Alexandria, a distance of 2872 nautical miles (n.m.). The distance from Plymouth to Lisbon was recorded as 770 n.m.; from Lisbon to Gibraltar as 302 n.m.; from Gibraltar to Malta as 980 n.m. What is the distance from Malta to Alexandria?

**5** (B). If a 5-ton truck can carry 1600 bricks when fully loaded and takes 17 full loads from a stack of 50,000 bricks, find (*a*) the number of bricks delivered, (*b*) the number remaining in the stack.

**6** (T). From the measurements given in fig. 2, of the dimensions of a screwed spindle, find the thickness of the packing gland.

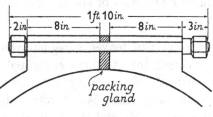

Fig. 2

**7** (A). In 1938 there were 41,100 acres under sugar-beet in the county of Norfolk. If the yield of beet worked out at 15 tons per acre, what was the total tonnage of sugar-beet raised in the county for that year?

**8** (Tr).  What would be the weekly mileage involved in a daily return run between Newbury and Bath if there is no Sunday service?

**9** (B).  If a lorry can carry 1600 bricks and four of such lorries are under contract to deliver 44,800 bricks to a builder, how many trips must each lorry make if they all make the same number?

**10** (A).  In eight successive weeks a poultry keeper collects the following numbers of eggs: 1484, 1491, 1466, 1501, 1492, 1533, 1538, 1577.  In this period he uses 226 eggs for his own household needs, and sells the rest.  How many dozen eggs does he sell?

**11** (T).  For the artificial lighting of a certain workshop it is estimated that 17,440 candle-power will be required.  How many lighting 'points' will be needed if lamps of 160 candle-power are to be used?

**12** (B).  A contractor for a drain-laying operation estimates that the work will require 23 men working for 13 days at 8 hr. a day.  How many man-hours of labour does this represent in his estimate of costs?

**13** (Tr).  A delivery van makes a regular round trip once a day, for 6 days a week, from Cirencester through Malmesbury, Cricklade and Fairford.  If the mileage recorder, before setting out on Monday morning, reads 24,334 miles, what would you expect the reading to be at the end of the round on the following Saturday?

**14** (B).  The estimate for plastering a wall is £3. 6*s*.  If the wall is found to measure 24 sq.yd. in area, what is the price per square yard for the plastering?

**15** (A). The following table gives the total area of agricultural land (T.A.A.) in the five counties in S.W. England. The area not under cultivation, being mountain and heath (M. and H.), is also shown:

| County | T.A.A. (acres) | M. and H. (acres) |
| --- | --- | --- |
| Cornwall | 721,000 | 111,000 |
| Devon | 1,438,000 | 303,000 |
| Somerset | 891,000 | 92,000 |
| Wiltshire | 743,000 | 133,000 |
| Dorset | 490,000 | 61,000 |

Find (*a*) the total agricultural area in the five counties of S.W. England, (*b*) the total area under mountain and heath, (*c*) the total area under cultivation.

**16** (Tr). What would be the amount of petrol used in a return journey by lorry from Lechlade to Upavon via Wantage and Hungerford if its petrol consumption is 8 miles per gallon?

**17** (B). Best Marchioness slates are usually cut so that 1200 slates weigh 48 cwt. (*a*) How many of these slates would weigh 1 cwt.? (*b*) What weight of slates must be supported on a roof carrying 850 of these slates?

**18** (T). A box-making machine turning out fourteen finished boxes every 3 min. is operating for 8 hr. daily. If the weight of material required for each box is 1 lb., what weight of material, in tons, is used in a week of 5 working days?

[ 5 ]

**19** (Tr). A flying-boat sets out at 06.00 hr. on 15 June to fly from New York to Archangel in Russia. The route is given by the following table:

| From | To | Distance in nautical miles |
|---|---|---|
| New York | St John's | 1080 |
| St John's | Reykjavik | 1060 |
| Reykjavik | Murmansk | 1560 |
| Murmansk | Archangel | 300 |

If the flying-boat flies throughout at a ground speed of 160 knots, at what time should it be due at Archangel, allowing 8 hr. for landing and refuelling on the journey?

**20** (A). The following table shows the number of head of cattle, sheep and pigs for England and Wales in 1928 and in 1939:

|  | Cattle | Sheep | Pigs |
|---|---|---|---|
| England (1928) | 5,208,376 | 12,889,126 | 2,724,659 |
| Wales (1928) | 818,057 | 4,000,484 | 246,384 |
| England and Wales together (1939) | 6,770,000 | 17,986,000 | 3,515,000 |

(*a*) From this table find the respective totals of (i) cattle, (ii) sheep, (iii) pigs, for England and Wales together in 1928. (*b*) From the totals given for 1939 find and state the increase or decrease in each kind of livestock as compared with 1928.

**21** (B). A brick pillar has to be constructed, forty courses high with thirty-two bricks in each course. If each brick weighs

[ 6 ]

7 lb., which of three available lorries could be used to deliver the bricks in one load, a 30-cwt., a 3-ton or 4-ton lorry?

**22** (Tr). On a straight section of railroad the permanent way is laid with rails each 30 ft. long, and the steel in each rail weighs 84 lb. per yard length. Find (*a*) the total number of 30 ft. rails, (*b*) the total weight in tons of the rails, in one mile of straight single-track permanent way.

**23** (Tr). A 3-ton lorry is loaded with eighteen bundles of steel rods, each bundle weighing 2 cwt. How many kegs of white lead each weighing 56 lb. can be added to the load without overloading the lorry?

**24** (T). A certain type of rotary press can print, fold, count and stack 5400 copies of newsprint per hour. What is the smallest number of such machines that will be needed to print 500,000 copies of the paper, if each machine operates for 4 hr. non-stop?

# CHAPTER 2

# FRACTIONS

## Revision of Vulgar Fractions

A fraction is always a part of something and the number above the line is called the *numerator* and that below the line the *denominator*. It is the number below the line that decides the name of the fraction. For example, $\frac{3}{16}$ denotes three-sixteenths of one whole number (unity) and $\frac{5}{32}$ in. denotes a length which is five thirty-seconds of 1 in.

Any fraction may be multiplied both in the numerator and the denominator by the same number without changing its value. All that is changed is its name, i.e. $\frac{5}{16} = \frac{5 \times 4}{16 \times 4}$ (twenty sixty-fourths).

In the same way, any fraction may be divided both in the numerator and the denominator by the same number without altering its value. In fact, no fraction which can be so divided by any number should be left without the division being made.

The fraction is then said to be reduced to its *lowest terms*.

## Addition and Subtraction of Fractions

Fractions may be added to or subtracted from one another just as whole numbers are, except that before doing so it is necessary to bring all fractions to the same name.

*Example.* What is the value of $\frac{15}{16} + \frac{5}{8} - \frac{1}{2}$? By bringing them all to sixteenths we have

$$\frac{15}{16} + \frac{10}{16} - \frac{8}{16}.$$

[ 8 ]

This is usually written, for convenience, as

$$\frac{15+10-8}{16}=\frac{17}{16}.$$

In this case we have a fraction with its numerator greater than the denominator (called an *improper fraction*). By dividing by 16 we obtain the answer $1\frac{1}{16}$, which is called a *mixed number*.

When adding or subtracting mixed numbers, find the sum or difference of the whole number part first and then deal with the fractions.

*Example.* What is the value of

$$4\tfrac{3}{16}+2\tfrac{1}{4}-3\tfrac{7}{8}?$$

Taking the whole number part first, we have

$$6-3+\frac{3}{16}+\frac{1}{4}-\frac{7}{8}=3\,\frac{3+4-14}{16}=3\,\frac{7-14}{16}.$$

We cannot subtract 14 from 7 so we 'borrow' one of the three whole numbers (leaving 2) and write it as $\frac{16}{16}$.

So that, $\qquad 3\,\dfrac{7-14}{16}=2\,\dfrac{16+7-14}{16}=2\tfrac{9}{16}.$

In actual practical measurement, apart from scientific and microscopic measurement, only very few fractions are used and they are almost invariably half, quarters, eighths, sixteenths, thirty-seconds and sixty-fourths, and, in accurate machining, thousandths. Nuts and bolts measure $\frac{5}{8}$ in., $\frac{3}{4}$ in., $\frac{11}{32}$ in. and so on. Commodities, when sold by weight, are sold in their appropriate units (cwt., lb., etc.) and the fractions, if they are needed, are $\frac{1}{2}$ and $\frac{1}{4}$. No one buys $\frac{3}{11}$ lb. of nails.

In industry you would never be troubled with such problems as $\frac{13}{29}+\frac{11}{57}-\frac{2}{13}$ and all such types are omitted from these

[ 9 ]

exercises. The addition or subtraction of fractions in industrial work is usually a simple and straightforward operation.

## Multiplication and Division of Fractions

When two fractions have to be multiplied or divided the process is very simple. The need for this usually arises from having to alter the scale of a working drawing, or for some similar reason. For example, a drawing may have to be made $\frac{1}{8}$ full size and some of the dimensions of the full-scale object are in fractions.

The rule is: in multiplication always multiply the numerators together to get the new numerator, and the denominators to obtain the new denominator.

In division, invert the fraction by which you are dividing, and then multiply as above; cancelling when possible.

*Example.*
$$\frac{3}{16} \times \frac{3}{4} = \frac{9}{64}$$

and
$$\frac{5}{8} \div \frac{1}{4} = \frac{5}{\underset{2}{\cancel{8}}} \times \frac{\cancel{4}}{1} = \frac{5}{2} = 2\frac{1}{2}.$$

The last answer, $2\frac{1}{2}$, is perhaps at first sight rather unexpected. But if the problem is considered to mean 'how many quarters are there in five eighths?' or better 'how many times will two eighths divide into five eighths?' then the answer is seen to be correct.

When multiplying or dividing mixed numbers always change the mixed number into an improper fraction and proceed as before.

*Example.*
$$2\frac{3}{4} \times \frac{1}{16} = \frac{11}{4} \times \frac{1}{16} = \frac{11}{64}$$

and
$$2\frac{3}{4} \div \frac{1}{16} = \frac{11}{\cancel{4}} \times \frac{\overset{4}{\cancel{16}}}{1} = 44.$$

[ 10 ]

## Fractions of Quantities

It is often necessary to convert a measurement in one unit to its corresponding value in another unit.

For example, when estimating for a job requiring a number of lengths of material, say, 4 ft. 4½ in. long, it is much easier to convert the 4½ in. into a fraction of a foot rather than to bring the length to inches.

So that we wish to find what fraction are 4½ in. of 1 ft.

*Rule.* Place the quantity immediately *following* the word 'of' in the denominator, thus: $\overline{1 \text{ ft.}}$

Place the other quantity in the numerator: $\dfrac{4\frac{1}{2} \text{ in.}}{1 \text{ ft.}}$.

Bring both to the same unit, i.e. inches:

$$\frac{4\frac{1}{2} \text{ in.}}{1 \text{ ft.}} = \frac{4\frac{1}{2}}{12} = \frac{\overset{3}{\cancel{9}}}{2} \times \frac{1}{\underset{4}{\cancel{12}}} = \frac{3}{8}.$$

So that the above measurement is 4⅜ ft. Notice that the above rule holds good when the problem is differently worded: what fraction of 1 ft. are 4½ in.?

## EXERCISE 2

**1** (T). What would bolts of the following diameters measure if tested with a vernier micrometer reading $\frac{1}{1000}$ths of an inch? (a) ½ in., (b) ¼ in., (c) ⅜ in., (d) $\frac{5}{16}$ in.

**2** (A). Fig. 3 is a plan of a small vegetable garden divided into three equal sections.

(a) What fraction of the garden is under root crops? (b) What fraction is not under root crops? (c) If half of the seed

[ 11 ]

plot is planted with peas, what fraction of the whole garden is under peas? (*d*) If the green vegetable plot is planted with savoys, brussels sprouts and cauliflower, so that the same amount of ground is given up to each, what fraction of the whole garden is under savoys?

| Root crops | Seed plot | Green vegetables |
|---|---|---|

Fig. 3

**3** (Tr). Fig. 4 represents the dial of a speedometer.

(*a*) What speed is being recorded? (*b*) What is the maximum speed that the speedometer will measure. (*c*) What fraction of top speed is the speedometer now recording?

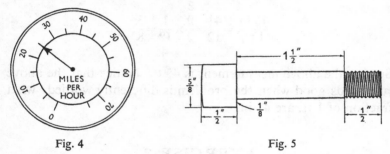

Fig. 4                    Fig. 5

**4** (T). Fig. 5 is a drawing of a steel bolt, with dimensions shown.

(*a*) What is the overall length of the bolt? (*b*) What fraction of the overall length is the threaded length? (*c*) What is the thickness of the bolt?

**5** (T). The resistance that a conductor offers to the passage of an electric current is measured in *ohms*. If a number of wires

[ 12 ]

are joined together in *series*, i.e. end to end so as to form one long wire, the total resistance of such a wire is the sum of the separate resistances. What is the resistance of the wires in series as shown in fig. 6?

$3\frac{1}{4}$ ohms    $2\frac{5}{8}$ ohms    $4\frac{11}{16}$ ohms

Fig. 6

**6 (T).** The reciprocal of a number is obtained by placing that number in the denominator of a fraction with 1 as the numerator, e.g. the reciprocal of 4 is $\frac{1}{4}$. The reciprocal of a fraction, or a mixed number written as an improper fraction, is found by turning the fraction upside down (inverting).

*Example.* The reciprocal of $\frac{1}{8}$ is $\frac{8}{1} = 8$. The reciprocal of $3\frac{3}{16}$ (i.e. $\frac{51}{16}$) is $\frac{16}{51}$.

When resistances are joined in *parallel*, i.e. side by side, the final resistance is *not* the sum of the resistances as in series. It is obtained by adding the *reciprocals* of the various resistances and inverting the final answer. Thus, the final resistance of 2 ohms and 4 ohms in parallel is found in this way:

Sum of the reciprocals $= \frac{1}{2} + \frac{1}{4} = \frac{3}{4}$.

∴ The final resistance $= \frac{4}{3} = 1\frac{1}{3}$ ohms.

What are the final resistances of the wires in parallel as shown in fig. 7 (*a, b, c*)?

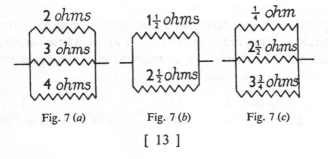

Fig. 7 (*a*)      Fig. 7 (*b*)      Fig. 7 (*c*)

[ 13 ]

**7 (B).** Standard bricks measure 9 in. by $4\frac{3}{8}$ in. by $2\frac{11}{16}$ in., but, allowing for the mortar used in jointing, they take up a space equal to 9 in. by $4\frac{1}{2}$ in. by 3 in. in ordinary building work.

They are usually laid so that each course of bricks is 3 in. high.

Fig. 8 shows a part of a 9 in. wall (i.e. a wall 9 in. thick), three courses high, laid in a pattern known as 'English bond'. The first course consists of bricks laid lengthways in pairs (called

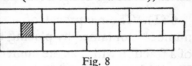

Fig. 8

*stretchers*). The second course consists of single bricks laid from front to back of the wall (called *headers*). The shaded portion is a wide gap of mortar (a *heading course*) which is inserted to break the line of joints, otherwise there would be a straight vertical line of mortar at the end of each stretcher throughout the whole wall, which would seriously weaken its strength.

(*a*) How many bricks are used in building the section of 9 in. wall as shown? (*b*) What fraction of the wall consists of headers? (*c*) It is necessary to rebuild a section of 9 in. wall laid in English bond. The section measures 3 ft. long by 1 ft. high. How many bricks will you need for the job? (*d*) A wall is 3 ft. high laid in English bond. How much of this height is made up of mortar?

**8 (G).** Every map has a scale of distance marked upon it and this is sometimes given in the form of a fraction (called the *representative fraction* of the map).

Suppose that two aerodromes are shown on a map and, by measurement, they are found to be 1 in. apart. These aerodromes are known to be actually 2 miles apart. What is the representative fraction (R.F.) of the map?

$$\text{R.F.} = \frac{\text{map distance}}{\text{ground distance}} = \frac{1 \text{ in.}}{2 \text{ miles}} = \frac{1}{2 \times 1760 \times 36} = \frac{1}{126,720}.$$

This means that: a map distance of 1 in. represents an actual distance of 126,720 in. and a map distance of 1 ft. represents an actual distance of 126,720 ft.

The measurement on a map of the distance between a light-vessel moored off the coast, and a harbour entrance is 4 in. Their actual distance apart is 1 mile. What is the R.F. of the map?

**9** (B). A line of tiles, acting as a skirting, has to be laid along the base of a wall measuring 22 ft. 3 in. in length. The tiles are $8\frac{1}{2}$ in. square. What is the least number of tiles that you would need for the job?

**10** (Tr). An aircraft has a tank capacity of 120 gal. of fuel of which $\frac{1}{6}$ must be kept as an emergency reserve. At cruising speed the aircraft uses 35 gal. per hr.

(*a*) What fraction is the hourly consumption of the available fuel? (*b*) How many hours cruising can the aircraft undertake with this fuel load?

**11** (T). A fitter has to turn a spindle of diameter $\frac{17}{32}$ in. He has two metal rods from which to choose. One measures $\frac{5}{8}$ in. diameter and the other $\frac{9}{16}$ in. Which should he use?

**12** (B). Fig. 9 represents a roof truss. *AB* is called the 'tie rod' and its length is known as the *span* of the roof. *CD* is the 'king bolt' and its length is called the *rise* of the roof. *AC* and *BC* are 'rafters', and *DE* and *DF* are 'struts'. How much timber (in feet) is there in

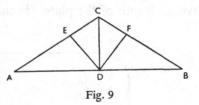

Fig. 9

eight of such roof trusses if the measurements of each truss are these: half span = 8 ft. 9 in., rise = 4 ft. $7\frac{1}{2}$ in., each rafter = 9 ft. 9 in., each strut = 4 ft. $4\frac{1}{2}$ in.?

[ 15 ]

**13** (T). Fig. 10 represents a metal pipe in section. The internal diameter (*d*), measured with inside calipers, is $2\frac{13}{16}$ in. The external diameter (*D*), measured with outside calipers is $3\frac{1}{4}$ in. What is the thickness of the metal?

**14** (T). A cruiser's measurements are: length 636 ft., breadth 75 ft., depth 45 ft. What would be the corresponding measurements, in inches, of a model constructed to a scale of $\frac{1}{12}$ in. = 1 ft.?

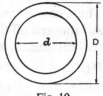

Fig. 10

**15** (G). In a tidal estuary the depth of water at half-tide is found to be 24 ft. When the tide is $\frac{2}{3}$ full the depth is 27 ft. What is the depth of water at (*a*) high tide, (*b*) low tide?

**16** (A). The weight of saleable beef obtained by the slaughter of a bullock is calculated as being $\frac{4}{7}$ of the live weight of the animal. What would a butcher estimate as the total weight of saleable beef from two bullocks, one weighing $10\frac{1}{2}$ cwt. and the other $9\frac{5}{8}$ cwt.?

**17** (T). Fig. 11 is a drawing of a metal link plate with two circular borings. The various dimensions are given. Find (*a*) the overall length of the plate, (*b*) the length *AB*.

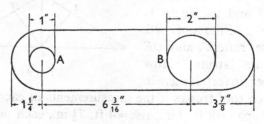

Fig. 11

[ 16 ]

**18** (B). The safe load that can be carried on hard stock bricks set in mortar is $4\frac{3}{4}$ tons per sq.ft. What is the greatest load that should be placed on a pillar of such bricks if the base of the pillar measures $9\frac{1}{2}$ sq.ft.?

**19** (Tr). A lorry starts a journey with the petrol indicator showing $8\frac{1}{4}$ gal. of petrol in the tank. The mileage-recorder reading is 12,394 miles. At the end of the trip the petrol indicator shows $5\frac{1}{2}$ gal. and the mileage recorder 12,438 miles.

(*a*) What is the petrol consumption of the lorry in miles per gal.? (*b*) What was the average speed of the lorry in miles per hr. if the journey took $2\frac{1}{2}$ hr.?

**20** (T). Electrical leads are usually made of insulated copper wire, in strands, of standard thickness (or *gauge* as it is called). The thinner the wire the greater is the resistance. If insulated copper wire of standard gauge 19/20 offers $1\frac{1}{4}$ ohms resistance per 1000 yd., what would be the resistance of 4 miles of this wire?

**21** (T). In fig. 12 the wire $AB$ is an electrical feeder carrying current via a junction box at $B$ to three lighting circuits. The total consumption of current (measured in *amperes*) in the lighting circuits must come from the feeder. How many amperes are flowing in the feeder?

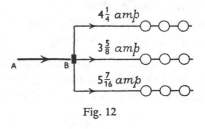

Fig. 12

**22** (T). Fig. 13 shows the elevation and plan of a screwed bolt, washer and nut. The sizes, as will be seen, depend upon the value of $D$ (the thickness of the bolt). By reading the drawings, find the following measurements, if the thickness of the bolt is

[ 17 ]

# FRACTIONS

$\frac{9}{16}$ in.: (*a*) thickness of the nut, (*b*) thickness of the square bolt head, (*c*) length of one side of the square head, (*d*) width of the nut across the flats, (*e*) the size of a suitable washer.

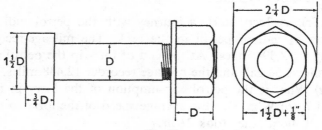

Fig. 13

**23 (T).** Fig. 14 represents two long electric mains supplying current to a number of filament lamps. The current consumption of each lamp is $\frac{3}{16}$ amp. The total current being supplied to the lamps is measured by the *ammeter* (*A*). If the ammeter needle shows a consumption of $9\frac{3}{8}$ amp., how many lamps are 'burning'?

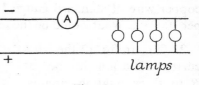

Fig. 14

**24 (T).** In the circuit below, fig. 15, a 5 amp. fuse is shown in both the positive and negative electric mains. A fuse is a short length of wire of low melting-point, designed to melt, and

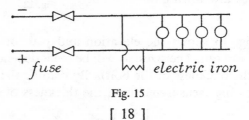

Fig. 15

[ 18 ]

thereby break the circuit, if the current consumption rises above a certain amount (in this case 5 amp.).

In the circuit are a number of lamps each consuming $\frac{1}{8}$ amp., and an electric iron which takes $2\frac{3}{4}$ amp. when in use. What is the greatest number of lamps that can be switched on at the same time without 'blowing' the fuse, (a) when the iron is not in use, (b) when the iron is in use?

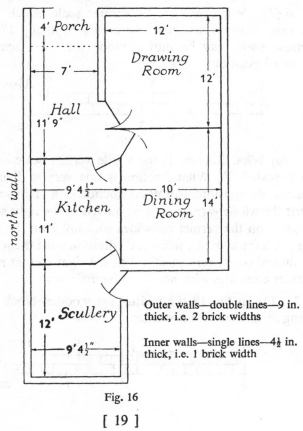

Fig. 16

[ 19 ]

**25** (B). From fig. 16, the ground plan of a house (p. 19), determine (*a*) the overall width, (*b*) the overall length, of the building at ground-level. In all the 9 in. walls a damp course, consisting of bituminous sheeting, is laid between the brick courses at a height of 6 in. above ground-level. This material is supplied, in the correct width, in rolls each containing 24 ft. (*c*) How many rolls are needed for the job, assuming no wastage?

**26** (T). A simple type of vernier scale which is used for making accurate measurement is shown in fig. 17, where the vernier scale may be moved along the main scale into any required position.

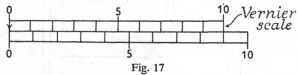

Fig. 17

(*a*) What fraction is the whole vernier scale of the whole main scale? (*b*) What fraction is one vernier division of one main-scale division? (*c*) What fraction of a main-scale division must the whole vernier scale be moved to the right so that no. 1 division on the vernier coincides with no. 1 on the main scale? (*d*) What fraction of a main-scale division must the whole vernier be moved (from zero position) to the right so that no. 7 on the vernier coincides with no. 7 on the main scale?

**27** (T). In fig. 18 the width of a wooden block (shaded) is being measured with a vernier.

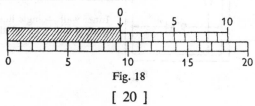

Fig. 18

[ 20 ]

(*a*) Which unit on the vernier coincides with a unit on the main scale? (*b*) What is the measurement of the width of the block in main-scale units?

**28** (T). In fig. 19 the main scale is 1 in. in length and it is divided into eight equal divisions. The vernier is divided also into eight equal divisions.

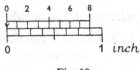

(*a*) What fraction, in this case, is one small vernier division of 1 in.? (*b*) What fraction of 1 in. must the

Fig. 19

vernier be moved from zero to the right so that no. 5 division on the vernier coincides with the 5th small division on the inch scale?

**29** (T). In fig. 20 this vernier is being used to measure the diameter of a steel rod. From the reading of the vernier, what is the diameter of the rod?

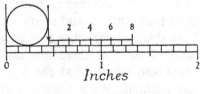

Fig. 20

**30** (T). A sample of Admiralty gunmetal is found to consist of $2\frac{1}{4}$ lb. of copper, 1 lb. 5 oz. of zinc and 6 oz. of tin. What weight of copper would be needed to make one ton of this gunmetal?

# CHAPTER 3

# DECIMALS

## Revision of Decimals

A decimal is a fraction, although it is not written as such, in which the denominator is always 10, 100, 1000, etc., according to the number of *useful* figures in the decimal. By useful figures we mean those that determine the value of the decimal fraction. In the decimal ·375 (which as a vulgar fraction may be written $\frac{375}{1000}$) each figure obviously has a meaning. The fraction may be written in this form:

$$\tfrac{300}{1000} + \tfrac{70}{1000} + \tfrac{5}{1000} = \tfrac{3}{10} + \tfrac{7}{100} + \tfrac{5}{1000}.$$

So that, in a decimal, the figure immediately following the decimal point indicates the number of tenths in the decimal. The second figure shows the number of hundredths; the third figure the number of thousandths, and so on.

In the case of the decimal ·005, the two noughts *are* useful figures because they show that there are no tenths and no hundredths and they also denote that the 5 is the number of thousandths in the decimal. As a vulgar fraction, therefore, ·005 is $\frac{5}{1000}$.

Any number of noughts *following* the last figure of a decimal do not alter its value. They are not useful figures.

For example, ·251 has the same value as ·25100.

In practical measurement and in the use of machine tools there is generally an allowance (or *tolerance* as it is called) in the accuracy of measurement or machining that is called for. So that a tolerance of one thousandth of an inch would mean that the measurement must be made to the nearest ·001 of an inch. When making practical measurements there is little object in carrying a decimal fraction (as is sometimes done) beyond the third decimal place, since there is little chance of there being any reasonable method of measuring the difference.

In the following examples, therefore, the answer should not be worked beyond the third decimal place (see *division of decimals*).

## Significant Figures

The answer to a decimal problem is sometimes asked for to a certain number of *significant figures*. Although the number of 0's immediately following the decimal point are useful figures they do not, without the rest of the decimal, give any indication of its actual value.

For example, in the decimal ·00214, the two 0's denote that there are no tenths and no hundredths. It is the figures 214 that tell us the value of the decimal, and these are called significant figures. The decimal ·00214 has, therefore, three significant figures.

It is always the practice, to avoid any error, to insert a 0 *before* the decimal point of any decimal with no whole number, so that ·214 is always written 0·214 and ·005 is written 0·005.

## Addition and Subtraction of Decimals

When adding or subtracting decimals the points must always be kept under one another, including that in the answer.

*Example.*  16·98 + 2·075 + 11·6 and 31·2 − 5·56.

$$
\begin{array}{ll}
16\cdot98 & 31\cdot2 \\
2\cdot075 & 5\cdot56 \\
\underline{11\cdot6} & \overline{25\cdot64} \\
\overline{30\cdot655} &
\end{array}
$$

## Multiplication and Division of Decimals

*Example.* Multiply 3·14 by 1·09.

First multiply without points:

$$
\begin{array}{r}
314 \\
109 \\
\hline
2826 \\
3140 \\
\hline
34226 \\
\end{array}
$$

Count the total number of useful figures after the decimal points in *both* the quantities to be multiplied (in this case 4).

Counting from the right, count 4 figures and then insert the decimal point.

The answer, then, is 3·4226.

*Example.* Divide 3·795 by 2·66.

First convert the divisor into a whole number by moving the decimal point two places to the right, so that 2·66 becomes 266. Equalize this by moving the decimal point in the dividend the same number of places to the right, so that 3·795 becomes 379·5.

Now proceed as in ordinary division but place the decimal point in the answer *immediately* after bringing down the first figure in the decimal, thus:

$$
\begin{array}{r}
1\cdot4 \\
266\overline{\smash{\big)}\,379\cdot5} \\
266 \\
\hline
1135 \\
1064 \\
\hline
71 \\
\end{array}
$$

The division is incomplete, and may be carried as far as necessary by bringing down a succession of 0's from the decimal part of the dividend.

If the answer is required to the third decimal place, and the division is still not then complete, it should be carried to the fourth place.

If the fourth figure is less than 5 it is ignored; if it is 5 or over, then 1 is added to the third place to obtain the nearest correct answer to the third decimal place. Thus:

$$
\begin{array}{r}
1{\cdot}4266 \\
266\,\overline{)\,379{\cdot}5} \\
266 \\
\hline
1135 \\
1064 \\
\hline
710 \\
532 \\
\hline
1780 \\
1596 \\
\hline
1840 \\
1596 \\
\hline
\end{array}
$$

The answer, in this case, to the nearest third decimal place is 1·427.

## Conversion of Decimals to Vulgar Fractions and vice versa

*Example.* Convert 2·045 into a vulgar fraction.

Put the whole number in front of the fraction thus: 2—.

Place all the useful figures after the decimal point as the numerator of the fraction: $2\frac{045}{}$.

For the denominator place 1 for the decimal point and as many noughts as there are figures after the decimal point. So that the fraction becomes $2\frac{45}{1000} = 2\frac{9}{200}$.

## DECIMALS

*Example.* Express ⅜ as a decimal. The fraction ⅜ means 3 divided by 8. If we carry out this division we obtain the answer as a decimal, thus:

$$8 \underline{)\ 3{\cdot}000}$$
$$\phantom{8\ )\ }0{\cdot}375$$

## EXERCISE 3

**1** (A). Rainfall is usually measured in inches, and decimals of an inch. The following are the monthly totals from 1 Jan. to 31 Dec. 1945 in a certain farming district: 2·12, 3·05, 3·26, 1·5, 1·62, 0·9, 0·78, 0·68, 0·47, 1·91, 3·74, 4·03. Find (*a*) the total rainfall for 1945 in that district, (*b*) the difference between the greatest and least month's rain.

**2** (Tr). The measurement of atmospheric pressure is a matter of great importance to pilots engaged in air transport. It is usually measured by a barometer and recorded for convenience in terms of inches of mercury. The normal atmospheric pressure at sea-level for the British Isles is 29·92 in. of mercury. The highest recorded barometric reading ever recorded in these islands is 31·11 in. of mercury, and the lowest reading is 27·33 in.

How much is (*a*) the highest reading above the lowest, (*b*) the highest above normal, (*c*) the lowest below normal?

**3** (Tr). In meteorology, or weather forecasting, atmospheric pressure is more often expressed in *millibars* (mb.) instead of in inches of mercury. This is to facilitate international exchange of weather information, and in this case normal atmospheric pressure in this country is 1013·2 mb.

So that 29·92 in. of mercury represent 1013·2 mb.

Two seaplane stations *A* and *B* are 500 miles apart. The atmospheric pressure at *A* is 981·1 mb., and at *B* it is 965·3 mb. What fall in pressure is there, at sea-level, between *A* and *B*?

**4 (T).** The diagram in fig. 21 represents a metal plate with three circular drilled holes. From the dimensions given, find (*a*) the overall length of the plate in inches, (*b*) the measurements of *x* and *y*, if the radius *R* is 1·09 in. and the radius *r* is 0·68 in.

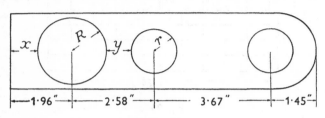

Fig. 21

**5 (Tr).** The pressure of the atmosphere, due to its weight, gradually decreases the higher we go and this decrease in pressure is fairly uniform for low altitudes up to 5000 or 6000 ft. Atmospheric pressure falls by 1 mb. for every 30 ft. increase in altitude above sea-level. If the atmospheric pressure is recorded as 932·4 mb. in an aircraft flying at 900 ft. above sea-level, what would be the pressure (*a*) at sea-level, (*b*) at an altitude of 2130 ft.?

**6 (T).** Fig. 22 represents four resistances in series. If the total resistance *R* is 12·12 ohms, what is the resistance of *r*?

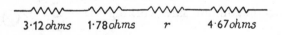

3·12 *ohms*  1·78 *ohms*  *r*  4·67 *ohms*

Fig. 22

**7 (B).** The unit of measurement for brickwork is the *standard rod*. A wall 272 sq.ft. in area and a brick and a half thick (i.e. 9 in. + 4½ in. = 13½ in.) contains a 'standard rod' of brick-work. Similarly, a wall with three times the area and half a brick thick (4½ in.) would also contain a standard rod.

The number of standard rods in a wall 13½ in. thick is found by multiplying the length of the wall in feet, by the height of the wall in feet, by 0·00368.

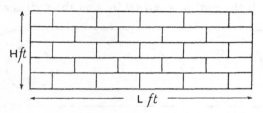

Fig. 23

Thus, in fig. 23, if the wall is a brick and a half thick then the number of standard rods of brickwork (S) is found from:

$$S = L \text{ ft.} \times H \text{ ft.} \times 0 \cdot 00368.$$

Find the number of standard rods in such a wall 120 ft. long by 6 ft. high.

**8** (T). The safe load which should be lifted by an iron chain should not be greater than 0·167 of the weight which will break the chain (called the *breaking strain*).

If the breaking strain of an iron chain with links 1 in. in diameter is 25·6 tons, what is its maximum safe working load in tons?

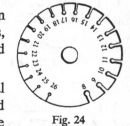

**9** (T). Fig. 24 represents an Imperial Standard Wire Gauge (I.W.G.) which is used to measure the diameter of a wire or the

Fig. 24

[ 28 ]

thickness of a metal sheet. The gauge of a particular wire or sheet is the number of the slot into which it fits exactly.

The table below gives the actual size, in decimals of an inch, of gauges numbered 15 to 17:

| I.W.G. no. | Inches |
|------------|--------|
| 15 | 0·072 |
| 16 | 0·064 |
| 17 | 0·056 |

A length of copper wire, of I.W.G. 16, is insulated by material which is 1·21 times as thick as the wire. What is the overall diameter of the insulated wire to the nearest third decimal place?

**10** (B). From the ground plan, fig. 16, p. 19, determine the number of standard rods of brickwork, above ground-level, in the north wall when it has reached a height of 10 ft. during construction.

The formula for a 9 in. wall is:

$$S = L \text{ ft.} \times H \text{ ft.} \times 0·00248.$$

**11** (A). A churn full of new milk contained 16½ gal. and weighed 209 lb. If the weight of the empty churn was 38½ lb., find, to the second decimal place, the weight of a gallon of new milk.

**12** (G). An English sprinter in a continental race won the 100 metres in 11·3 sec. If 1 metre = 1·094 yd., what was his speed in yd. per sec., to the nearest first decimal place?

**13** (B). The weight of ½ in. thick roofing slates is reckoned as 7·5 lb. per sq.ft. What weight of slates would have to be supported on a roof measuring 3600 sq.ft.? Give your answer in cwt. and decimals of a cwt.

**14** (T). The *pitch* of a screwed thread is the distance that the screw moves forward with one complete turn of the screw. Find, to the nearest thousandth of an inch, the pitch of screws having (*a*) 8, (*b*) 11, (*c*) 14 threads to the inch.

**15** (Tr). The scale of a road map is 1·75 in. = 1 mile. What is the ground distance between two places measuring, on the map, 6·27 in. apart?

**16** (Tr). A vessel, on speed trials, is found to consume 4 tons of fuel oil per hr. with her engines developing 5600 horse-power. What is her oil consumption in lb. per unit horse-power per hr.?

**17** (Tr). A passenger locomotive, running alone at 40 miles per hr., consumed 21 lb. of coal per mile. The same engine pulling eight passenger coaches at the same speed used 1 ton 6 cwt. of coal on a 30-mile run. Find, to the nearest first decimal place, the coal consumption in lb. per passenger coach per mile.

**18** (T). Fig. 25 represents a headed tapered cotter used for keying a wheel on to shafting. The size of the cotter depends

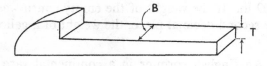

Fig. 25

upon the diameter of the shafting used, and if the diameter of the shafting is $D$ in., then:

$$B = \frac{D}{4} + \frac{1}{8} \text{ in.} \quad \text{and} \quad T = \frac{D}{10} + \frac{1}{16} \text{ in.}$$

If the diameter of the shafting is 2¼ in., find (*a*) the correct breadth, (*b*) the correct thickness (to the nearest hundredth of an inch) of the cotter.

**19** (G). The total capacity, or volume, in cubic feet, of all the enclosed spaces in a ship, divided by 100 is called the *gross tonnage* of the ship. (It is assumed that 100 ft. of space are equivalent to 1 ton weight.) So that a ship with 125,000 cu.ft. of enclosed space has a gross tonnage of 1250 tons.

The *nett, or register, tonnage*, upon which harbour and canal dues, etc., are usually paid, is calculated from the gross tonnage by making an allowance for crew's quarters and machinery space. In merchant ships the nett, or register, tonnage is usually 0·62 of the gross tonnage, and in warships it is generally assumed to be 0·45 of the gross tonnage.

A ship's *displacement tonnage* is the total weight of the ship and her maximum cargo, or what is the same thing, the weight of water that such a ship displaces when afloat.

If the total capacity of all enclosed spaces of a merchant ship is 725,000 cu.ft., what is her nett, register, tonnage?

**20** (G). During the refitting of a battleship her gross tonnage was increased from 27,250 tons to 31,310 tons. What was the corresponding increase in her nett, register, tonnage?

**21** (G). From the following details estimate to the nearest ton the nett, register, tonnage of an aircraft carrier: (*a*) displacement tonnage 26,500, (*b*) gross tonnage 0·54 of displacement tonnage, (*c*) crew space 0·05 of gross tonnage, (*d*) machinery, boilers, hangars, bunkers, etc., 0·46 of gross tonnage.

**22** (T). By using the reciprocal method (Ex. 2, 6 (T)), estimate the actual resistance of wires in parallel as in (a) fig. 26a, (b) fig. 26b.

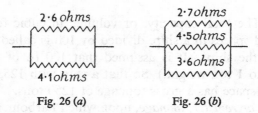

Fig. 26 (a)       Fig. 26 (b)

**23** (T). What is the measurement, in main-scale units, of the wood block in fig. 27?

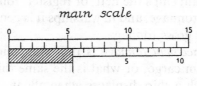

Fig. 27

**24** (T). Fig. 28 shows steel vernier calipers, graduated in inches, being used to measure the diameter of a half-crown. What is the reading?

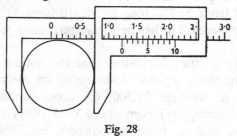

Fig. 28

# CHAPTER 4

# AVERAGES

A very simple, and useful, arithmetical process is 'finding the average'.

Suppose that in a workshop the following numbers of geared spindles of a certain type were used in five consecutive working weeks: 211, 230, 209, 241, 240.

The average weekly consumption of these could then be determined by adding together the five quantities and dividing by 5, i.e. $\frac{1131}{5} = 226$ (to the nearest whole number). It should be remembered that for an average to serve any useful purpose the quantities, from which the average is found, must be *reliable* quantities. It would be no use to include in the above figures any one which was abnormal, by reason of holidays, shortage of staff, breakdown or any other cause. When we have obtained a useful and representative average we can use this figure as a forecast of future needs, and base our forward orders for stock, over a period, with reasonable confidence.

It is also important to remember that the time interval when taking an average must be the same for each quantity.

For example, the average speed of a ship which steams 12 knots for 3 hr., 17 knots for 4 hr. and 16 knots for 5 hr. is found by calculating the *total* distance travelled and dividing by the *total* time taken.

Thus:

Distance travelled in 3 hr. at 12 knots = 36 n.m.
,,       ,,       4 hr. ,, 17 knots = 68 n.m.
,,       ,,       5 hr. ,, 16 knots = 80 n.m.

Therefore the distance travelled is $(36+68+80)$ n.m. in $(3+4+5)$ hr.

I.e. 184 n.m. are travelled in 12 hr. So that the average speed is $\frac{184}{12} = 15\frac{1}{3}$ knots.

# EXERCISE 4

**1** (A). During six consecutive years the average annual rainfall in a certain farming district was 29·9 in. In the first five years the actual rainfall was recorded as: 29·2 in., 27·4 in., 30·5 in., 28·8 in. and 31·2 in. What was the actual rainfall in the sixth year?

**2** (T). A circular steel rod is tested for uniformity of diameter with a micrometer screw gauge at half-inch intervals. The following readings are obtained: 0·573 in., 0·569 in., 0·570 in., 0·571 in., 0·573 in., 0·573 in., 0·572 in. What is the average diameter of the rod to the nearest third decimal place?

**3** (B). To make a special paint, 5 gal. of white lead paint, costing 12s. 6d. per gal., 1½ gal. of best black, costing 26s. per gal., and ½ gal. of thinners, costing 7s. per gal., are mixed together. What is the cost per gal. of the special paint?

**4** (T). In a dockyard the following numbers of 6 in. bolts were used on six consecutive normal working days: 350, 280, 410, 296, 376 and 346. If the job is estimated to take a further 78 working days, what would be the expected requirements for these bolts to the nearest thousand?

**5** (Tr). A motor car, under test, showed the following performance:

(*a*) during a 3 hr. run 78 miles were covered for a petrol consumption of 3½ gal.;

[ 34 ]

(*b*) during a 2 hr. run 56 miles were covered for a petrol consumption of 2¼ gal.;

(*c*) during a 1½ hr. run 45 miles were covered for a petrol consumption of 1¾ gal.

Find, to the nearest first decimal place, (1) the average speed during the tests, (2) the average mileage per gallon of fuel.

**6 (A).** A plot of ground, under potatoes, consisted of 270 rows. To estimate the approximate weight of the crop, three rows were lifted and weighed; one row near the middle and one near each end. The yield of these rows was 3 cwt. 74 lb., 2 cwt. 105 lb. and 3 cwt. 17 lb. respectively. Estimate the expected yield of the whole crop to the nearest ton.

**7 (T).** Two contractors are employed to do some excavation and carting. The first contractor removes 12 loads averaging 30 cwt. each. The second contractor removes 15 loads averaging 3 tons each. If the total cost of the job is £100. 16*s*., what is (*a*) the average cost per ton for excavation and carting, (*b*) the share each contractor receives?

**8 (T).** A battery of accumulators was used to supply electrical energy in a small workshop. The demand, in current consumption, made upon the battery during 12 hr. was as follows:

From 6 a.m. until noon, a current of 15 amp.

„    noon    „   4 p.m.      „      12 amp.

„    4 p.m.   „   6 p.m.      „      10 amp.

What was the average current taken from the battery for the whole period?

**9 (B).** Three bricklayers, each with his own labourer, are engaged on building a wall. The first man lays 32 bricks in 40 min., the second man lays 26 bricks in 30 min. and the third man lays 22 bricks in 24 min. Estimate the probable number of bricks laid if the three men work steadily for a 4 hr. shift.

**10** (Tr). The following are the recordings of high and low water for three normal days at Plymouth:

H.W.  14·8 ft.; 15·0 ft.; 14·0 ft.; 14·1 ft.; 13·3 ft.; 13·5 ft.
L.W.  0·4 ft.; 0·7 ft.; 1·6 ft.; 2·3 ft.; 2·9 ft.; 3·3 ft.

What is the average rise of tide (i.e. the difference in height between high and low water) at Plymouth?

**11** (T). In a motor-car factory the output of completed cars for five successive normal working days was: 36, 34, 38, 27 and 30. If each car needs 8 yd. of fabric for upholstery, estimate the approximate number of rolls of this material (each roll containing 66 yd.) that will be required for a month of 24 full working days.

**12** (G). The average weight of the oarsmen in a racing 'eight' was 11 st. 6½ lb. The average weight of the whole crew, including the coxswain, was 11 st. 1 lb. What was the weight of the coxswain?

# CHAPTER 5

# RATIO

A ratio is used to denote a relationship between any two or more quantities of the same kind.

For example, brass is an alloy made of a 'mixture' of two metals, copper and zinc. Many different varieties of brass may be made by altering the *ratio* of copper to zinc in the alloy.

Muntz metal is a brass containing 3 parts by weight of copper and 2 parts by weight of zinc (making a total of 5 parts by weight of the Muntz metal).

This relationship may be written thus, in the case of Muntz metal; wt. of copper : wt. of zinc $= 3 : 2$, or $\dfrac{\text{wt. of copper}}{\text{wt. of zinc}} = \dfrac{3}{2}$.

When two quantities bear a known ratio to each other, and the value of one is known, then the value of the other may be easily found.

*Example.* How much copper must be alloyed to 12 lb. of zinc to make Muntz metal?

From the above ratio we have:

$$\frac{\text{wt. of copper}}{12 \text{ lb.}} = \frac{3}{2}.$$

I.e. wt. of copper $= \dfrac{12 \times 3}{2}$ lb. $= 18$ lb.

It is quite common to have mixtures of more than two substances. A valuable artificial manure for agricultural purposes consists of a mixture of (*a*) nitrate of soda (3 parts by weight), (*b*) superphosphate (4 parts by weight), (*c*) Kainit (6 parts by

[ 37 ]

weight). This may be written $a:b:c = 3:4:6$, so that the total number of parts by weight is $3+4+6 = 13$.

Therefore, whatever weight of this artificial fertilizer we mix, $\frac{3}{13}$ of the total weight must be nitrate of soda, $\frac{4}{13}$ of the total weight must be superphosphate and $\frac{6}{13}$ must be Kainit.

## EXERCISE 5

1 (B). When estimating the cost of plastering, the following ratio is often used:

Labour costs : cost of material $= 3:4$.

If the total cost of material for plastering the ceiling and walls of a room was £7. 2s. 8d., find (a) the cost of the labour, (b) the total cost of the job.

2 (T). Tinsmith's solder is an alloy of lead and tin, in the proportion 7 parts by weight of lead to 3 parts by weight of tin. If tin is seven times the price of lead, what is the ratio of the cost of lead to tin in this solder?

3 (A). A load of $1\frac{1}{2}$ cwt. of artificial manure is required, consisting of a mixture of sulphate of ammonia, superphosphate and potash in the ratio $2:3:1$ by weight. What weight of each is there in the mixture?

4 (Tr). In engine design the ratio $\dfrac{\text{wt. of engine in lb.}}{\text{horse-power developed}}$ is known as the *weight-power ratio* and should, particularly in aero-engines, be as small as possible.

The first primitive aero-engine used by the Wright brothers weighed 210 lb. and developed only 30 H.P., so that its weight-power ratio was $\frac{210}{30} = 7$.

# RATIO

Find the weight-power ratio for each of the following engines:

|  | | Wt. in lb. | H.P. |
|---|---|---|---|
| (a) | Armstrong Siddeley Panther | 980 | 560 |
| (b) | Rolls Royce Merlin Mk. 600 | 1760 | 1,650 |
| (c) | Rolls Royce Nene (Jet) | 1600 | 12,000 |

Find the H.P. of the following engines:

|  | | Wt. in lb. | W/P ratio |
|---|---|---|---|
| (d) | Bristol Pegasus III | 1060 | 1·758 |
| (e) | Bristol Mercury | 1015 | 1·676 |

Find the weight of the following engines:

|  | | H.P. | W/P ratio |
|---|---|---|---|
| (f) | Parsons Marine Turbine | 18,000 | 2·65 |
| (g) | Bristol Pegasus II | 580 | 1·827 |

5 (T). For all circles, irrespective of size, the ratio $\dfrac{\text{circumference}}{\text{diameter}}$ has a constant value, denoted by the Greek letter $\pi$. The ratio is not an exact whole number and an approximation is $3\frac{1}{7}$, or more nearly 3·142.

A roller, 15 in. in diameter is being used as a friction drive for a smaller roller (fig. 29). If it is required that the smaller roller shall revolve five times as fast as the larger one, what must be the diameter of the small roller?

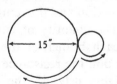

Fig. 29

6 (B). In a new road three houses have been built with frontages to the road of 36 ft., 42 ft. and 50 ft. respectively. If the total cost of road charges is £166. 8s., what is (a) the cost per foot frontage, (b) the amount payable by each householder?

[ 39 ]

**7** (T). Fig. 30 shows how the cutting blades of a 'garden' type of lawn mower are made to revolve. The inner casing of the driving wheel $D$ is toothed and these teeth engage with similar teeth on the spur wheel $S$. The spur wheel itself is fixed to the axle of the cutting blades. If there are 72 teeth on the driver and 16 on the spur wheel, find (*a*) the number of times each cutting blade revolves when the machine is pushed forward a distance equal to one revolution of the driver, (*b*) what must be the diameter of the spur wheel if the diameter of the driver is 9 in.?

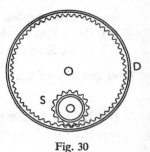

Fig. 30

**8** (A). A field of wheat, at harvest, gave a yield of 48 bushels of grain to the acre, together with 36 cwt. of straw per acre. A bushel of wheat weighs 63 lb. What was the weight ratio of wheat to straw for this field?

**9** (T). When an electric current arrives at a point $A$ in a circuit (fig. 31) where an alternative path is offered, then part of the current will traverse one section and the remainder the other. The quantity of current passing in each branch of the circuit depends upon the resistance of each branch. Obviously, if the resistances are equal, then half of the current will pass into one branch and half into the other.

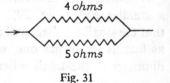

Fig. 31

Suppose, as in fig. 31, that the resistances are different, namely 4 ohms and 5 ohms. Then more current will traverse the 4-ohms resistance than the 5 ohms, and actually the ratio is the opposite of the resistance ratio.

Thus, in this case, the resistance ratio is $4:5$, so that the current ratio is $5:4$, and $\frac{5}{9}$ of the current will flow in the 4-ohms resistance and $\frac{4}{9}$ in the 5-ohms resistance.

What is the current ratio, between $D$ and $E$ (fig. 32), when (*a*) the hinged stud $A$ is switched to $B$, and (*b*) when switched to $C$?

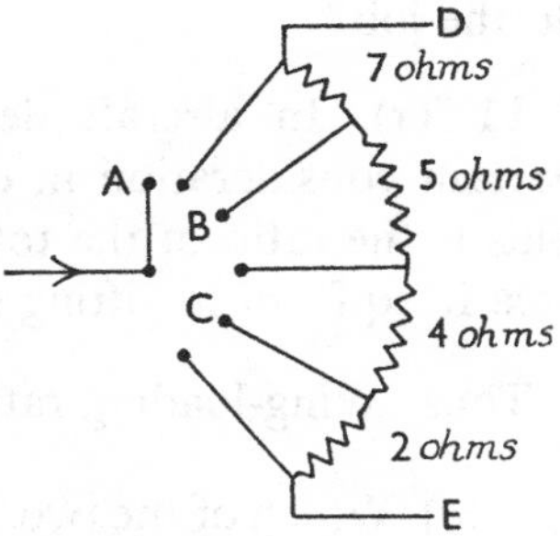

**10** (B). All ropes have what is known as a *breaking strain*, which is the smallest weight that will break the rope.

Fig. 32

It is most essential, for reasons of safety, to know what is the breaking strain of any rope or tackle that is lifting heavy weights. The safety margin that is allowed for use with any rope is very much less than the breaking strain, and the maximum *safe working load*, as it is called, is expressed as a ratio of the breaking strain.

Thus, the ratio $\dfrac{\text{safe working load}}{\text{breaking strain}} = \dfrac{1}{6}$ for wire ropes,

,, $= \dfrac{3}{8}$ for dry, hempen rope.

So that, if the breaking strain of any rope is known, then the safe working load is easily calculated. When rope is wet its safe working load is still further reduced by $\frac{1}{4}$.

(*a*) What is the safe working load of a wire rope of breaking strain 2 tons 5 cwt.?

(b) What is the safe working load of a dry hemp rope of breaking strain 6·4 cwt.?

(c) What would be the safe working load of rope (b) if wet?

(d) It is required, on a wet day, to raise a stone block weighing 4½ cwt. to the top of the outside of a building. To what breaking strain should the rope be tested to be sure that it will do the job?

**11** (Tr). In aircraft designing *wing-loading ratio* is an important consideration in deciding the stability of the machine. This is the ratio of the total weight in lb. of the aircraft to the area in sq.ft. of its lifting surface.

Thus: Wing-loading ratio $= \dfrac{\text{weight of aircraft in lb.}}{\text{area of lifting surface in sq.ft.}}$.

(a) Which of the two following aircraft has the higher wing-loading ratio and by how much? (work to the nearest first decimal place):

|  | Gross weight | Wing area |
|---|---|---|
| (1) Avro York | 65,000 lb. | 1297 sq.ft. |
| (2) Avro Tudor I | 76,000 lb. | 1421 sq.ft. |

(b) Approximately how many times the lifting power, in lb. per sq.ft., has the Avro Tudor I than Wright's aircraft of 1903, which had a weight of 925 lb. and a wing area of 600 sq.ft.?

**12** (T). A very important consideration in the distribution of electrical power throughout the country is the question of the most economical way in which to convey it from one point to another. At the power station the current when generated is usually an alternating current of high ampere strength driven by a low voltage. By 'alternating' we mean a current which changes its direction, backwards and forwards, and varies its strength many times a second (usually 50 times, called the 'cycle'). If transmitted in this form it would lead, by reason of heating and corrosion in the conducting wires, to much loss of energy and deterioration of equipment, so that it is much more economical to pass the electrical energy in the form of a current of low ampere strength driven by a high voltage.

The method by which the current is converted is by instruments known as *transformers*.

The essential principle of the transformer is very simple and consists of a core of soft iron 'stampings' on which are wound two separate series of windings, the *primary* and the *secondary*. The primary consists of a comparatively few coils of heavy insulated copper wire wound on the core and the secondary contains a large number of turns of fine insulated copper wire wound on the primary.

These two coils are quite independent and may be represented in diagrammatic form as in fig. 33.

The object is to 'transform' a powerful alternating current of high ampere strength, driven by a low voltage, into a weak current of low ampere strength, driven by a high voltage. Since it is the voltage that is of major importance in this operation, such a transformer is called a *step-up* transformer, as the voltage is stepped up in intensity.

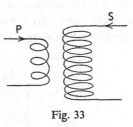

Fig. 33

The ratio $\dfrac{\text{number of turns in secondary}}{\text{number of turns in primary}}$ is called the *transformer ratio*, so that if there are 2000 turns in the secondary and 10 in the primary, the transformer ratio is 200 : 1.

[ 43 ]

In a step-up transformer the voltage is stepped up by this ratio and the ampere strength is stepped down by the same ratio.

In a *step-down* transformer the operation is reversed.

So that it is the usual practice to employ a step-up transformer at the generating end and a step-down at the consumer's end.

From the following, calculate (1) the transformer ratio, (2) the secondary voltage:

|     | Turns of primary | Turns of secondary | Primary voltage |
| --- | --- | --- | --- |
| (a) | 30 | 3600 | 15 |
| (b) | 24 | 6000 | 10 |
| (c) | 32 | 1280 | 16 |

**13 (T).** It is required to transform a current of 240 volts pressure to one with a line pressure of 11,280 volts.

(a) What transformer ratio is necessary?

(b) If there are 1410 turns on the secondary, how many are needed in the primary?

**14 (T).** It is required to transform an alternating current which is 100 amp. and 200 volts at the generator to a line pressure of 8000 volts and then to deliver it to the consumer at 250 volts. The circuit in diagrammatic form, is shown in fig. 34.

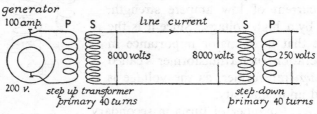

Fig. 34

(a) What is the number of turns necessary in the secondary of the step-up transformer?

(b) What is the line current?

(c) What is the step-down ratio at the consumer's end?

(d) What is the current strength supplied to the consumer?

(e) How many secondary turns are necessary in the step-down transformer?

**15** (T).   In the secondary of the step-up transformer in fig. 35

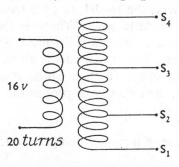

16 v

20 *turns*

Fig. 35

we have a number of separate points $S_1$, $S_2$, $S_3$ and $S_4$. Power may be drawn from any two of these four points. The total number of secondary turns is 6000.

Between $S_1$ and $S_2$ the transformer ratio is 50 : 1,

„    $S_1$  „  $S_3$     „           „       150 : 1.

How many secondary turns are there between (a) $S_1$ and $S_2$, (b) $S_1$ and $S_3$, (c) $S_1$ and $S_4$, (d) $S_2$ and $S_3$, (e) $S_2$ and $S_4$, (f) $S_3$ and $S_4$? (g) What is the secondary voltage when drawing from $S_3$ and $S_4$? (h) What is the transformer ratio between $S_2$ and $S_4$?

## CHAPTER 6

# PROPORTION

If we wish to make a model or a pattern of any object, such as a piece of machinery, the model, to be a true imitation of the original, must be made to *scale*. Whatever the scale (or ratio) chosen, for example 1 : 8, this ratio must be observed in all linear measurements. Then the model is a true scale model and all parts are in *proportion*. In other words, if we have a number of ratios all equal, then we have a proportion.

## Direct Proportion

Consider the case of a machine manufacturing some finished article at a known, steady rate of output. Then the machine will deliver three times as many finished articles in 3 hr. as it does in 1 hr. So that for any increase or decrease in time of operation there is a corresponding increase or decrease in output.

Such conditions constitute what is called *direct proportion*.

*Example* 1. If a machine turns out 5 gross of screws in $6\frac{1}{4}$ min., what is the output of the machine per hr.?

*Unit Method.* In $6\frac{1}{4}$ min. the machine turns out 5 gross.

Therefore in 1 min. „ „ $\dfrac{5}{6\frac{1}{4}}$ gross

and in 60 min. „ „ $\dfrac{60 \times 5}{6\frac{1}{4}}$ gross

$$= \frac{\overset{12}{\cancel{60}} \times \cancel{5} \times 4}{\underset{\cancel{5}}{\cancel{25}}} = 48 \text{ gross.}$$

*Fractional Method.* The greater the time taken, the greater the output. Therefore the new output will be $\frac{60}{6\frac{1}{4}}$ times the old output

$$= \frac{60}{6\frac{1}{4}} \times 5 = 48 \text{ gross.}$$

*Example* 2. A vessel loads 360 tons of grain in $2\frac{1}{2}$ hr. How long should it take to load 540 tons?

*Unit Method.* 360 tons are loaded in $2\frac{1}{2}$ hr.

Therefore      1 ton is     ,,     $\dfrac{2\frac{1}{2}}{360}$ hr.

and      540 tons are     ,,     $\dfrac{540 \times 2\frac{1}{2}}{360}$ hr.

$$= \frac{\overset{3}{\cancel{540}} \times 5}{\underset{2}{\cancel{360}} \times 2} = \frac{15}{4} = 3\frac{3}{4} \text{ hr.}$$

*Fractional Method.* More grain will take longer to load, so that the new time is $\frac{540}{360}$ of the old time

$$= \frac{540}{360} \times 2\frac{1}{2} = 3\frac{3}{4} \text{ hr.}$$

# EXERCISE 6

**1** (B). A 'mechanical' shovel can dig 160 yd. of trench for pipe-laying in 8 hr. How long should it take this machine to dig a similar trench half a mile long?

**2** (T). A certain electric cable offers a resistance of 4·9 ohms per mile. Find the resistance offered by the following lengths of the same cable: (*a*) 650 yd., (*b*) 1200 yd.

## PROPORTION

**3** (G). If 25 rupees, in India, are worth 36 shillings, what would you expect to get by changing 48 rupees into English money? (Answer to the nearest penny.)

**4** (A). The Ministry of Agriculture recommend that a certain type of artificial fertilizer should be spread so that 1 cwt. covers 720 sq.yd. of ground. What weight of this fertilizer would be needed to manure 540 sq.yd.?

**5** (Tr). From a transport aircraft in flight, at an altitude of 5000 ft., a photograph is taken of objects on the ground, and two of these objects on the photographic print are found to be 2 in. apart. The photographic plate in the camera is 6 in. behind the lens.

If the following proportion holds good:

$$\frac{\text{distance from lens to plate}}{\text{distance from lens to ground}} = \frac{\text{measurement on photograph}}{\text{measurement on ground}},$$

find how far apart these two objects actually are. (Answer to nearest yard.)

**6** (B). If a pull of 84 lb. on a certain lifting tackle during building operations is able to raise a weight of 756 lb., what weight could be raised, on the same tackle, with a pull of $\frac{1}{2}$ cwt.?

**7** (Tr). If a ship's log recorded a run of 182 nautical miles in 12 hr. 20 min., what was her average speed in knots?

**8** (T). A model of a Spitfire Mk. VIII is to be made $\frac{3}{32}$ full size. What are the corresponding measurements in inches of the model if the wingspan of the actual aircraft is 40 ft. and its length 31 ft.?

[ 48 ]

**9** (B).  Fig. 36 shows a roof truss with a span (*BC*) of 21 ft. and a rise (*AD*) of 9 ft. What is the length of the horizontal strap *EF* placed 5 ft. below the ridge (*A*)?

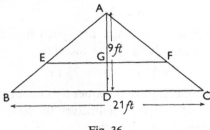

Fig. 36

**10** (T).  A motor attached by pulley to shafting operates a drilling machine. When the motor is revolving at 400 revolutions per min. (r.p.m.) the speed of the drill is 160 r.p.m. What is the drill speed when the motor is operating at its maximum of 1500 r.p.m.?

## Inverse Proportion

In some cases of proportion we find that as one quantity increases so the other quantity correspondingly decreases. For example, the *greater* the speed of a vehicle the *less* time is taken for the same distance covered.

The *greater* the thickness of a wire the *less* its resistance to the passage of an electric current.

Such cases are examples of *inverse proportion*.

*Example* 1.  If a train travels from London to Bristol in 3 hr. at a speed of 40 m.p.h., how long would the same journey take at a speed of 48 m.p.h.?

*Unit Method.* At 40 m.p.h. the train takes 3 hr. for the journey. At 1 m.p.h. the train would take $3 \times 40$ hr. for the journey.

At 48 m.p.h. the train would take $\dfrac{3 \times 40}{48}$ hr. for the journey.

$$= 2\tfrac{1}{2} \text{ hr.}$$

*Fractional Method.* The faster the speed the less time taken, therefore the new time is $\tfrac{40}{48}$ of the old time

$$= \tfrac{40}{48} \text{ of 3 hr.} = 2\tfrac{1}{2} \text{ hr.}$$

Problems in proportion are frequently encountered where we have a number of varying quantities to consider, involving possibly both direct and inverse proportion. These are best solved by the fractional method in separate stages.

*Example* 2. If 4 men engaged on drain laying can dig a trench 27 ft. long by 3 ft. wide by 4 ft. deep in 5 hr., how long should it take 5 men to dig a trench 54 ft. long by 2 ft. wide by 3 ft. deep with the same conditions?

First, the more men employed would reduce the time taken, so that the new time is $\tfrac{4}{5}$ of the old time

$$= \tfrac{4}{5} \text{ of 5 hr.}$$

The greater length of trench would take longer time, i.e. new time $= \tfrac{54}{27}$ of $\tfrac{4}{5}$ of 5 hr.

The less the width, the less the time, so, new time becomes $\tfrac{2}{3}$ of $\tfrac{54}{27}$ of $\tfrac{4}{5}$ of 5 hr.

The shallower the trench, the less the time taken, and the final time becomes $\dfrac{3}{4}$ of $\dfrac{2}{3}$ of $\dfrac{\overset{2}{54}}{27}$ of $\dfrac{4}{5}$ of 5 hr.

$$= 4 \text{ hr.}$$

## EXERCISE 7

**1** (T). One copper wire has 1·2 times the cross-sectional area of another. The thicker wire offers a resistance of 0·4 ohm. What would be the resistance of the same length of the thinner wire?

**2** (Tr). A lorry driver, leaving London at 8 a.m., has to deliver goods to Bristol, a distance of 120 miles, by 2 p.m. After averaging a speed of 20 m.p.h. for $1\frac{1}{2}$ hr. he is delayed for $\frac{1}{2}$ hr. with tyre trouble.

(*a*) What speed must he average for the remainder of the journey to arrive on time? (*b*) If he actually averages 25 m.p.h., at what time does he arrive?

**3** (B). The roof on a house of 16 ft. frontage is 10 ft. 6 in. from eaves to ridge (fig. 37). The total weight of slates on this roof is 35 cwt. What would be the weight of slate on a roof of 24 ft. frontage measuring 9 ft. from eaves to ridge?

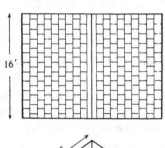

**4** (A). A recommended artificial manure should be spread at the rate of 3 oz. per sq.yd. and at this rate a plot of land 30 yd. long by 8 yd. wide would require 45 lb. of the fertilizer. What weight of this fertilizer would be needed to manure a plot 56 yd. long by 12 yd. wide with a distribution of $2\frac{1}{2}$ oz. per sq.yd.?

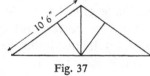

Fig. 37

**5** (T). A carbon-steel drill revolving at a speed of 50 r.p.m. takes 20 min. to drill through a wrought-iron plate $2\frac{1}{2}$ in. thick.

(*a*) How long would the same drill, at a speed of 60 r.p.m., take to drill through a 2¼ in. wrought-iron plate? (*b*) What decimal of an inch does the drill advance with each revolution?

**6 (B).** It is found that to deliver a stack of bricks, measuring 25 yd. long, 4 yd. wide and 6 ft. high, it is necessary for four 3-ton lorries to make 20 separate journeys each. How many journeys would each of five 3-ton lorries have to make to deliver the bricks in another stack measuring 27 yd. long, 5 yd. wide and 7 ft. 6 in. high?

**7 (A).** A herd of 40 dairy cows averages a milk yield of 2 gal. per cow every day. From this milk 32 lb. of butter is made daily. How much butter would you expect daily from a herd of 60 cows giving an average daily yield of 2¼ gal. per cow, assuming the proportion of fat in the milk is the same?

**8 (T).** The ratio $\dfrac{\text{engine speed in r.p.m.}}{\text{road wheel speed in r.p.m.}}$ is called the *gear ratio* of a motor vehicle.

A certain light lorry has three forward gears. When in first (or bottom) gear the ratio is 18 : 1, the second gear ratio is 10 : 1, and the third (or top) gear ratio is 6 : 1. When in first gear, with the engine revolving at 1000 r.p.m., the speed of the lorry on the level is 4·5 m.p.h.

Find the corresponding speeds (*a*) in second gear, (*b*) in top gear, (*c*) the maximum speed of the lorry if the maximum engine speed is 3600 r.p.m.

**9 (T).** Electrical resistance of a wire is in direct proportion to its length and in inverse proportion to its cross-sectional area.

If 360 yd. of copper wire of cross-sectional area 0·012 sq.in. offer an electrical resistance of 4·8 ohms, what will be the resistance of 450 yd. of copper wire with cross-sectional area of 0·008 sq.in.?

**10** (B). A 250 gal. hot water tank supplied for a house is found to be faulty. It measures 4 ft. 6 in. long, 3 ft. 6 in. wide and 2 ft. 6 in. deep. If it is replaced by one measuring 5 ft. long, 3 ft. wide and 2 ft. 3 in. deep, what, by proportion, is its capacity to the nearest gallon?

# CHAPTER 7

# PERCENTAGE

A percentage is merely a ratio in which the second term is always 100. It is not written as a ratio but has its own individual symbol.

Thus 80 per cent is written 80 % and indicates a ratio of 80 : 100 or $\frac{80}{100}$.

Any fraction, or decimal fraction, may be expressed as a percentage by multiplying it by 100.

So that, $\frac{5}{8}$, expressed as a percentage, is $(\frac{5}{8} \times 100)$ % $= 62\frac{1}{2}$ %, and 0·25 is $(0·25 \times 100)$ % $= 25$ %.

One advantage of expressing fractions or ratios as percentages is that their relative values may then be readily compared.

*Example.* A field $A$ of 9 acres and a field $B$ of 12 acres were both sown in wheat during a season and the yields were: $A$, 54 quarters and $B$, 75 quarters. The following season these fields were manured with artificial manures of different qualities and again seeded with wheat. The second season's crops were: $A$, 57 quarters, and $B$, 79 quarters. Which manure showed the better result?

The increase in yield for field $A = 57 - 54$
$$= 3 \text{ quarters.}$$

I.e. Percentage increase for field $A = \frac{3}{54} \times 100$
$$= 5·56 \text{ %.}$$

The increase in yield for field $B = 79 - 75$
$$= 4 \text{ quarters.}$$

I.e. Percentage increase for field $B = \frac{4}{75} \times 100$
$$= 5·33 \text{ %.}$$

So that it is at once evident that field *A* showed the better result.

*Note*. Notice particularly that all percentages showing comparisons of increase or decrease of output, etc., are *always* based upon the original figure and not the final one.

# EXERCISE 8

**1 (B).** In dressing 380 ft. of rough timber it is estimated that $2\frac{1}{2}$ % will be wastage. What length of dressed timber should be available for use?

**2 (B).** A job is estimated to require 250 ft. of dressed timber. How much rough timber (to the nearest foot) will be needed if the wastage is 5 %?

**3 (T).** Many metals, owing to their resistance, become hot when an electric current is passed through them and the hotter they get the more the resistance increases. The alloy known as 'manganin' is an exception to this general rule. Manganin is an alloy containing 84 % by weight of copper, 4 % by weight of nickel and the remainder manganese. Find the weight of each of these metals in a coil of bare manganin weighing $12\frac{1}{2}$ lb.

**4 (A).** A 5-acre field produced a harvest of 9·5 tons of clover hay. A sample of this hay was analysed and found to contain 2 % by weight of nitrogen.

It is assumed that this nitrogen has been taken from the soil. Find (*a*) the weight of nitrogen (to the nearest lb.) removed from the soil, per acre, by the growth of this hay crop. If it is decided to restore this shortage of nitrogen by nitrate of soda, which yields 17 lb. of nitrogen per cwt., find (*b*) the weight of nitrate of soda to distribute per acre to compensate for the clover crop.

**5 (B).** Concrete is made by mixing cement, fine aggregate (sand) and rough aggregate (gravel, shingle, broken brick, etc.) in certain proportions by volume. When dry there are considerable air spaces (called *voids*) between the particles of sand and still greater voids between the gravel, shingle or broken brick. The addition of water to the mixture brings about not only a chemical change by which the cement ultimately 'sets' into a hard mass, but when properly mixed the sand and cement fill the large voids between the rough aggregate while the finely ground cement fills the small voids between the particles of sand and so binds the whole solidly together.

For this reason, when mixing concrete, allowance has to be made for the consequent shrinkage, i.e. the reduction in volume, when the water is added to the dry mixture. This allowance is usually 35 %.

How much wet concrete, in cubic yards, will the following amounts make when properly wetted: cement, $\frac{1}{2}$ cu.yd.; sand, 1 cu.yd.; gravel, $2\frac{1}{2}$ cu.yd.?

**6 (B).** The footings of a retaining wall, which are to be made of concrete, measure $7\frac{3}{4}$ cu.yd. What is the volume of dry mixture that is needed (to the nearest cubic yard)?

**7 (B).** A job requires $3\frac{1}{4}$ cu.yd. of rough wet concrete. How much cement, sand and broken brick in the proportion 1 : 2 : 7 must be mixed for this purpose?

**8 (T).** The term 'work', when used scientifically has a special meaning, and work is only done when a weighted object is lifted, or a force is moved through a distance. (It is interesting to know that, technically speaking, no work is done by a man who, for example, holds a cwt. bag of cement on his back until he collapses from fatigue!)

The unit of work (the *foot-pound*) is that amount of work that is necessary to raise a mass of 1 lb. a distance of 1 ft.

So that to raise a gallon of water (which weighs 10 lb.) in a can which weighs 1 lb. through a height of 3 ft. needs:

$$11 \times 3 = 33 \text{ ft.-lb. of work.}$$

The rate of doing work is called *power* and the unit of power is the *horse-power* (H.P.) which is equal to 33,000 ft.-lb. of work done in 1 min. of time.

Therefore, if we raise the above can of water 3 ft. in 6 sec. we exert a power of $\dfrac{33}{33,000} \times \dfrac{60}{6} = \dfrac{1}{100}$th of a H.P.

In actual practice the amount of *useful* work that a machine can do is always less than the power that it develops.

Some of the power is lost or taken up by overcoming friction, developing unnecessary heat, etc.

And the ratio $\dfrac{\text{useful work obtained}}{\text{total work put in}}$ is called the *efficiency* of the machine.

If a 1 H.P. engine is capable of exerting only 26,500 ft.-lb. of work per min. its efficiency is $\dfrac{26,500}{33,000}$.

This ratio is always expressed as a percentage.

So that the efficiency of the above engine is

$$\left(\dfrac{26,500}{33,000} \times 100\right) \% = 80\% \text{ (approx.).}$$

A 400 H.P. colliery engine is capable of raising 5 tons of coal per min. from the bottom of a mine shaft 990 ft. deep. What is the efficiency of the engine?

**9** (T). A 25 H.P. fire engine can pump 39,600 gal. of water per hr. to a height of 100 ft. What is its efficiency?

**10** (T).  An engine rated at 3 H.P. is used on the hauling wire of a crane to raise a girder weighing 2¼ tons.  It lifts the girder through a distance of 24 ft. in 84 sec.

Find (*a*) the useful work done in ft.-lb. per min., (*b*) the work expended in overcoming friction, etc., (*c*) the efficiency of the engine.

**11** (Tr).  An accumulator, for the storage and release of electrical energy, is another type of machine from which, as always, we get less work out than we put in.

Its efficiency is measured by the ratio

$$\frac{\text{volts} \times \text{amperes} \times \text{hours of output}}{\text{volts} \times \text{amperes} \times \text{hours of charging}},$$

and this again is expressed as a percentage.

What is the efficiency of a 6-volt, 36 ampere-hour battery which requires 10 hr. for recharging with a 3 amp. current at 9 volts, when run down?

**12** (Tr).  A 12-volt motor-car battery has an output of 60 amp.-hr. and an efficiency of 75 %.  For how long must it remain on charge with a 4 amp. current at 10 volts in order to recharge it from an exhausted condition?

# CHAPTER 8

# GRAPHS

## Statistical Graphs. Plotting of Points

It is a common and convenient practice to illustrate statistical details or recorded facts in the form of 'graphs' (or more correctly in this case, 'charts'). A true graph is described later. For this purpose 'squared' paper is usually employed, ruled in squares of one-tenth of an inch.

Examine, for example, the following record of egg production from eight hens over a period of six calendar months:

| Month | Nov. | Dec. | Jan. | Feb. | Mar. | Apr. |
|-------|------|------|------|------|------|------|
| Eggs  | 60   | 104  | 131  | 151  | 164  | 144  |

To obtain a ready comparison of these totals we 'plot' them upon squared paper. First draw two straight lines $OA$, $OB$ at right angles, intersecting at $O$ (fig. 38). These are called the 'axes'. The horizontal axis $OA$ is used for the time factor (months) and the vertical axis $OB$ for the record of eggs collected. Any convenient scale of measurement is selected, say ten squares along the axis $OA$, to indicate a lapse of time of one calendar month (no allowance need be made for the variation in the number of days in different months) and one small square to represent 2 eggs along the axis $OB$, starting from some suitably low number—say 40 eggs—at $O$, called the 'point of origin'.

The first point, $P_1$, is plotted on the chart at the intersection of the vertical line through 'Nov.' on $OA$ and the horizontal line through '60 eggs' on $OB$. Point $P_2$ is similarly found from

'Dec.' on *OA* and '104 eggs' on *OB*, and so on until all the points are plotted. These points are usually joined together by

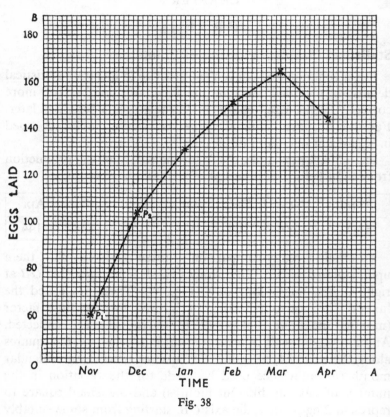

Fig. 38

straight lines which serve as a means to carry the eye readily from one point to the next.

Such a chart is merely a record of fact and has no value as a means of estimating intermediate results or for forecasting possible developments. The principal use of these charts is either to afford a comparison or to indicate a general trend of events.

## Continuous or Locus Graphs

These are true graphs and differ from statistical charts in that they *are* used mainly for the estimation of intermediate values or for forecasting.

When two quantities are known to be directly proportional to each other, then the graph illustrating this proportion is a straight line which usually passes through the point of origin.

Suppose that we wish to compare the gallonage of tanks of different dimensions measured in feet and inches. It is an easy matter to calculate the volume of the tank in cubic feet (or fractions of a cubic foot) and by means of a suitable graph we can quickly read off the corresponding gallonage for any volume in cubic feet that we need.

All we require to know is the comparison for any one volume, since these quantities are in direct proportion.

In this case let us use the relationship:

10 cu.ft. are equal to 62·5 gal.

We know also that this graph must pass through the point of origin, since, if the volume in cubic feet is 'nil', the gallonage must obviously be 'nil'.

Draw the two axes at right angles, as before, and label the horizontal axis—Cubic Feet, and the vertical axis—Gallons. (Fig. 39.) The scales chosen should depend upon the values likely to be needed. For example, if the graph is required to show comparisons for volumes of not more than 20 cu.ft. then the scale of cubic feet should be large. If we want comparisons up to, say, 200 cu.ft., the scale must be small.

[ 61 ]

Let us assume that we do not want to exceed 50 cu.ft. The most convenient scale in this case is: 1 small square = 1 cu.ft.

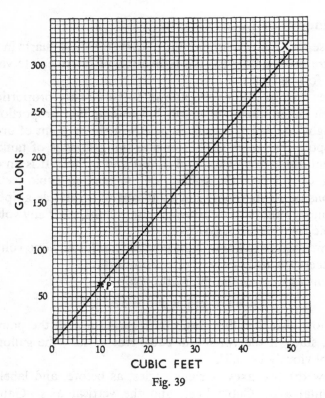

Fig. 39

The scale of gallons, in the same way, may be taken as: 1 small square = 5 gal.

Plot the known point $P$ (10 cu.ft. = 62·5 gal.) and draw the straight line $OX$ through this point and the origin.

Then $OX$ is the required graph from which as many equivalent capacities in gallons, as necessary, may be estimated for volumes up to 50 cu.ft. (or vice versa).

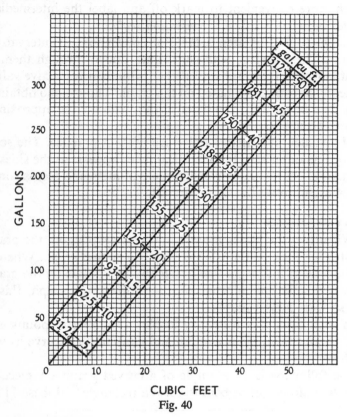

CUBIC FEET

Fig. 40

## Conversion Scales

A graph such as fig. 39 is often used to determine a full range of comparative values and is, in fact, employed as a comparative scale.

[ 63 ]

In such a case, in order to overcome the tedious monotony of 'running' the eye along and down the squared paper, it is much more convenient to mark off and label the intermediate values on what is called a *conversion scale*.

In fig. 40 will be seen a table of conversions at intervals of 5 cu.ft. These points, with short lines drawn through them at right angles to the graph are labelled with their respective values as nearly as can be estimated on the graph. The scale so obtained may then be used for rapid conversion with a close approximation to accuracy.

Many such conversion scales are used in practice. The scale of conversion of statute to nautical miles appears at the sides of all marine charts. A conversion scale of metres to feet is sometimes useful to a pilot in an aircraft.

## Curved Graphs

When two quantities are *not* in direct proportion, the graph illustrating their relationship is usually a curved graph. Whereas in the case of a straight-line graph any two points on the graph provided sufficient information to complete the graph, this is not so with curved graphs.

In such cases it is necessary to plot a number of points and then to join them by as smooth and continuous a curve as will fit them.

The following is an example of a curved graph. A piece of wire 24 in. long can be bent to form a rectangle with base 11 in.

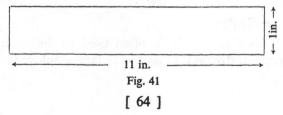

11 in.

1 in.

Fig. 41

and height 1 in. (fig. 41). It would then enclose an area of
$11 \times 1 = 11$ sq.in.

The wire could also be bent to form all the rectangles with
measurements as shown and enclosing the corresponding areas.

| Base (in.) | 11 | 10 | 9 | 8 | 7 | 6 | 5 | 4 | 3 | 2 | 1 |
|---|---|---|---|---|---|---|---|---|---|---|---|
| Height (in.) | 1 | 2 | 3 | 4 | 5 | 6 | 7 | 8 | 9 | 10 | 11 |
| Area (sq.in.) | 11 | 20 | 27 | 32 | 35 | 36 | 35 | 32 | 27 | 20 | 11 |

If we plot the graph connecting the relationship between base
and area for all these rectangles we obtain the curve shown in
fig. 42.

An examination of this graph shows us that the greatest area
we can enclose is 36 sq.in. Any increase in the length of the base
beyond 6 in. results in a diminution of area.

This point, $X$ on the graph, is called the turning point where
the area is a maximum. We can also read from the graph, that
to enclose an area of 25 sq.in. we must make the base approximately 2·75 in. long (point $P$) or 9·25 in. (point $P_1$).

## EXERCISE 9

**1** (A). The following details show the weight of a pig at
weekly intervals, over a period of nine weeks:

| | May | | | | | June | | | | July |
|---|---|---|---|---|---|---|---|---|---|---|
| Date | 1 | 8 | 15 | 22 | 29 | 5 | 12 | 19 | 26 | 3 |
| Weight (lb.) | $24\frac{1}{2}$ | $26\frac{1}{2}$ | 28 | 30 | 33 | 36 | $38\frac{1}{2}$ | 41 | 44 | 48 |

Plot a weight chart to illustrate these particulars.

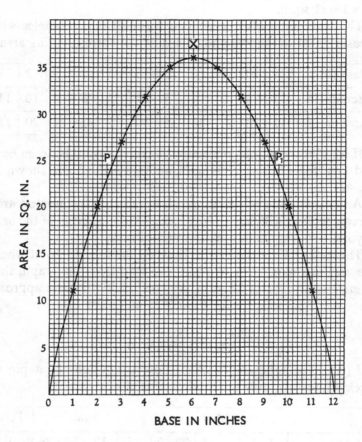

Fig. 42

**2** (Tr). Show, by means of a chart, a comparison of the daily oil-fuel consumption of a ship from the following details taken at successive noons:

| July | 3 | 4 | 5 | 6 | 7 | 8 | 9 | 10 |
|------|---|---|---|---|---|---|---|----|
| Fuel (tons) | — | 48 | 46 | 50 | 52 | 48 | 44 | 49 |

**3** (A). Plot a temperature chart for a manured mushroom bed from the following details taken at 8 a.m. on successive days: 62° F., 84° F., 97° F., 115° F., 112° F., 103° F., 84° F., 83° F.

**4** (T). The weight displacement of a ship is its total weight (including cargo, bunkers, etc.) when afloat. If a ship's weight displacement at noon on 8 Feb. was 4328 tons, plot, from the following tabulated details, a chart of the ship's daily weight displacement:

| Feb. | 8 | 9 | 10 | 11 | 12 | 13 | 14 | 15 |
|------|---|---|----|----|----|----|----|----|
| Fuel consumed (tons) | — | 42 | 46 | 42 | 48 | 46 | 47 | 42 |
| Cargo discharged (tons) | — | 130 | — | 50 | — | 120 | — | 180 |
| Cargo shipped (tons) | — | — | — | 210 | — | 80 | — | — |

**5** (B). When papering walls, which measure 7 ft. 6 in. to 8 ft. from skirting to picture-rail or cornice, an estimate of the number of rolls of paper required is shown in the following table. The perimeter of the room is the horizontal distance right round the four walls including doors, windows and fire grate.

| Perimeter (ft.) | 30 | 38 | 46 | 54 | 62 | 72 | 82 | 90 | 98 |
|-----------------|----|----|----|----|----|----|----|----|----|
| Rolls of paper | 4 | 5 | 6 | 7 | 8 | 9 | 10 | 11 | 12 |

Draw a conversion scale of the above and from it estimate the number of rolls required for a room 13 ft. 6 in. long by 11 ft. 9 in. wide and 7 ft. 9 in. high.

**6 (Tr).** Plot the graph of the following table representing the distance ($D$) in feet through which an object falls in air in $T$ seconds of time:

| Time (sec.) | 3 | 5 | 6 | 8 | 10 |
|---|---|---|---|---|---|
| Distance (ft.) | 150 | 350 | 450 | 700 | 1000 |

From this graph determine approximately how far a man falls before his parachute operates if he jumps from an aircraft, waits 5 sec. before pulling his rip cord and the parachute then takes $1\frac{1}{2}$ sec. to open.

**7 (T).** A discharged accumulator with a voltage of 1·9 volts is placed on charge. The following table shows how the voltage rises with the time of charging with a steady current of 6 amp.

| Time (hr.) | 0 | $\frac{1}{4}$ | 1 | 2 | 3 | 4 | 5 | $5\frac{1}{2}$ | 6 | 7 | 8 |
|---|---|---|---|---|---|---|---|---|---|---|---|
| Voltage of accumulator | 1·9 | 2 | 2·06 | 2·07 | 2·08 | 2·09 | 2·1 | 2·1 | 2·2 | 2·4 | 2·5 |

Plot the graph showing the relationship between time of charging and the voltage of the accumulator.

Assuming that the graph, if read backwards, indicates the discharge rate of the accumulator, (a) what are the limits of voltage between which the main discharge takes place? (b) What happens to the voltage in the early stages of discharge? (c) What is the output in amp.-hr. of the accumulator if its efficiency is 75 %?

**8 (T).** A certain type of fuse wire used in electric circuits is an alloy of lead and tin in the ratio 2 : 1. The following table indicates the current strength of the circuit at which the fuse wire will melt for different thicknesses of the fuse wire:

| Diam. of fuse wire (in.) | 0·01 | 0·02 | 0·03 | 0·04 | 0·05 |
|---|---|---|---|---|---|
| Current (amp.) | 1·3 | 3·7 | 6·8 | 10·5 | 14·4 |

Plot the graph showing this relationship and from it determine (a) the current strength that will fuse a wire of diameter 0·035 in., (b) the diameter of the fuse wire necessary for a circuit in which the current strength is not to rise above 5 amp.

| Speed (m.p.h.) | 15 | 20 | 25 | 30 | 35 | 40 |
|---|---|---|---|---|---|---|
| Miles per gal. | 11·2 | 11·6 | 11·9 | 11·95 | 11·8 | 11·4 |

**9 (Tr).** From the above table, representing the number of miles per gal. of petrol covered by a lorry at different speeds, plot the graph.

Find, from the graph, (a) at what speed the mileage per gal. attains a maximum, and (b) if the tank capacity is 12 gal., find how far the lorry can travel at the most economical fuel consumption.

**10 (B).** The following particulars apply to a pulley system consisting of two double-sheaved pulleys used for lifting weights. The *velocity ratio* of the system (denoted by $V$) in this case is 4 : 1. This means that 4 ft. of rope must be hauled in in order to lift the load a distance of 1 ft. In other words the effort ($P$) must move four times as fast as the load ($W$) whatever the value of the latter may be.

[ 69 ]

In the table below the effort $P$ and the load $W$ are both expressed in lb. The efficiency of the system at any load is

$$\left(\frac{W}{P \times V} \times 100\right) \%.$$

| Load $W$ | $5\frac{1}{2}$ | 12 | 17 | 25 | 30 | 37 | 45 |
|----------|------|------|------|------|------|------|------|
| Effort $P$ | 3 | $5\frac{1}{2}$ | 7 | 10 | 12 | $14\frac{1}{2}$ | 17 |
| Efficiency | 45·8 | 55·5 | 61 | 62·5 | 62·8 | 63·8 | 66·2 |

From the above details plot (i) an effort-load graph, (ii) an effort-efficiency graph.

From these graphs find (a) the effort necessary to raise a 40 lb. load and the efficiency of the system at that load, (b) the load that can be raised with an effort of 13 lb. and the efficiency of the system at this load.

[ 70 ]

## CHAPTER 9

# GEOMETRY

## Scales and Scale Drawing

A number of problems that would require considerable calculation to obtain an exact answer may be solved with sufficient accuracy by geometrical methods.

In such cases drawings are made to scale and measurements carefully taken with geometrical instruments, which include ruler, set squares, compasses and sometimes a circular protractor for measuring and constructing angles.

Accuracy with the use of any of these instruments is largely a matter of care and practice. Always work with a sharply pointed pencil and draw fine, thin, clean lines so that errors in measurement are reduced to a minimum.

A simple geometrical construction often found necessary when drawing to scale is the division of a line into a number of equal parts.

In fig. 43, suppose that the line *AB*, 2·35 in. long, is to be divided into five equal parts. From *A*, at any convenient angle

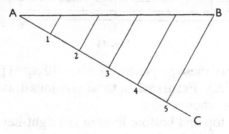

Fig. 43

[ 71 ]

(about 30°), draw a straight line *AC*. With dividers or compasses mark off five equal divisions from *A* along *AC*. Number these 1 to 5 as shown. Join no. 5 to *B* and then from 4, 3, 2 and 1 draw lines parallel to *B*5 cutting the line *AB*. The line *AB* is thus divided into five equal parts.

## The Diagonal Scale

With the ordinary ruler it is not easy to draw a straight line of exact length, or to measure with accuracy its length when drawn.

For this purpose we often employ a special measuring device called a *diagonal scale*. This can be purchased or one may be drawn upon good quality card and its accuracy will be reliable.

The construction is as follows: with a ruler and set square draw a rectangle 4 in. long by 1 in. high, and draw perpendiculars at 1 in. intervals as shown in fig. 44. This will divide the rectangle into four 1 in. squares.

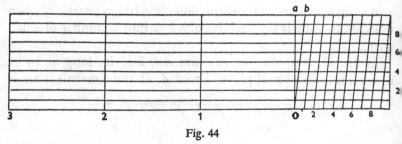

Fig. 44

Divide one of these perpendiculars into 10 equal parts (i.e. each division is $\frac{1}{10}$ in.). From each of these divisions draw lines parallel to the base, as shown.

Divide the top and bottom lines of the right-hand square also into $\frac{1}{10}$ths of an inch. Join these 'diagonally' as shown in the

[ 72 ]

diagram and label the scale with the figures indicated. This diagonal scale will then measure accurately to $\frac{1}{100}$ in. (0·01 in.).

This will be readily understood if we consider the small triangle 0$ab$. The distance from the 3 in. mark to the point 0, along the bottom horizontal line, is obviously 3 in. The distance from the 3 in. mark to the point $b$, along the top horizontal line is 3·1 in., since $ab = 0·1$ in.

We know that the ten horizontal lines are spaced at intervals of $\frac{1}{10}$ in. so that the length from the 3 in. mark along the first horizontal to the intersection with the diagonal 0$b$ is

$$3 + (\tfrac{1}{10} \text{ of } 0·1) \text{ in.} = 3·01 \text{ in.}$$

The distance from the 3 in. mark along the number 2 horizontal to its intersection with the diagonal 0$b$ is

$$3 + (\tfrac{2}{10} \text{ of } 0·1) \text{ in.} = 3·02 \text{ in.},$$

and so on.

To measure a length equal to 2·64 in., place the point of the dividers on the point of intersection of the number 6 'diagonal' with the number 4 horizontal. Open the dividers along the number 4 horizontal to the 2 in. mark. The distance thus measured is 2·64 in.

To measure a line of unknown length, first determine its approximate length with a ruler. If greater than 4 in. (since this is the maximum measurement we can make directly with the scale as drawn in fig. 44), mark off a whole number of inches, by the ruler, so that the remainder is less than 4 in. Suppose, for example, that the line is more than 8 in. long, then mark off 5 in. by ruler, leaving between 3 and 4 in. to be measured. Open the dividers accurately over this remaining length and with the point on the 3 in. mark of the diagonal scale determine which intersection agrees with the measured length. If the

[ 73 ]

agreement is along the number 7 horizontal at the intersection with the number 5 diagonal, then the length measured is 3·57 in., and the length of the line in question is $5 + 3·57 = 8·57$ in.

The following exercises, which include simple examples in the reading and construction of workshop drawings, afford practice in the use of geometrical instruments and accurate measurement by means of the diagonal scale.

## EXERCISE 10

**1** (A). The plan herewith (fig. 45) represents a small agricultural area. The scale is attached. Find, by scale measurement (*a*) the distance from the bridge *A* to cross-roads at *B*, (*b*) how far it is by the dotted footpath from *C* to *D*.

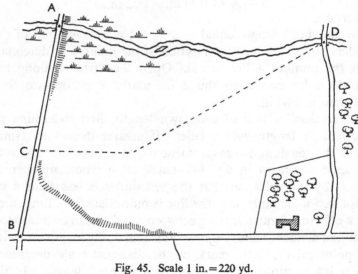

Fig. 45. Scale 1 in. = 220 yd.

[ 74 ]

**2** (G).  A buoy, marking a fairway, is anchored to the sea-bed by a mooring chain 25 ft. long.  Find, by drawing to a suitable scale, the radius of 'swing' of the buoy if the depth of water is (*a*) 21 ft., (*b*) 15 ft.

**3** (T).  A wireless mast 65 ft. long is to be erected with 5 ft. buried in the ground.  It is to be stayed by three equally spaced wire stays.  Each stay is fixed to a collar 20 ft. from the top of the mast and also to a stake in the ground 30 ft. from the foot of the mast.  Find, by drawing the mast and one stay to a scale of 10 ft. = 1 in., the total length of wire required for the stays, allowing 2 ft. extra to each stay for splicing.

**4** (B).  Fig. 46 represents a derrick which has been rigged for hauling roofing slates.  Draw this derrick to a scale of 10 ft. = 1 in. and from your drawing find (*a*) the length of the derrick pendant, (*b*) the distance from the fall to the foot of the derrick, if the derrick pendant is made 30 ft. long.

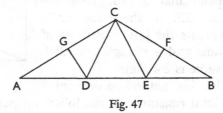

Fig. 46          Fig. 47

**5** (T).  The diagram (fig. 47) represents a steel roof truss in which *AD* = *DC* = *CE* = *EB*.  The struts *DG* and *EF* are at right angles to *AC* and *BC* respectively.

Draw to a scale of 1 in. = 5 ft. (using the diagonal scale for necessary measurements) a similar steel truss in which the span is 30 ft. and the angles *CAB* and *CBA* are each 30°. What is (*a*) the length of each strut, (*b*) the rise of the roof? (Answer to the nearest inch.)

**6** (T). The actual resistance (called the joint resistance) of two resistances in parallel (fig. 48) may be found geometrically (as well as mathematically as in Ex. 2 and 3) and it affords good exercise in careful drawing and measurement to compare results.

Suppose that the two resistances are 6 ohms and 4 ohms.

Draw any base line *AB* of convenient length and choose a suitable scale to represent 1 ohm resistance.

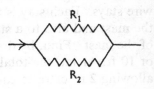

Fig. 48

At *A* draw a perpendicular *AC* six units in length and at *B* another perpendicular *BD* four units in length. (Fig. 49.)

Join *AD* and *BC*. From the point of intersection *E* draw *EF* perpendicular to *AB*. Then the length *EF*, to the same scale, represents the joint resistance of 6 ohms and 4 ohms in parallel. Its value is 2·4 ohms.

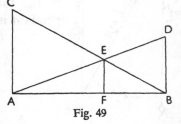

Fig. 49

Use this method to determine the joint resistance of the following pairs of wires in parallel: (*a*) 5 ohms and 8 ohms, (*b*) 2·4 ohms and 3·6 ohms, (*c*) 3·3 ohms and 6·7 ohms. Answer to second decimal place. Check your answers by the 'reciprocal' method in Ex. 3.

**7** (T). Steel pylons are used to support the insulators which carry high-tension electrical cables over long distances. Fig. 50

represents one side of a typical pylon the base of which is a square with 10 ft. sides. The four corner members (*AB* and *CD* are two of these) are of heavy angled steel and slope inwards so that the square at the top has 4 ft. sides. The length of each corner member above ground is 24 ft. Horizontal and diagonal struts render the structure rigid. The horizontal struts are at 8 ft. intervals.

Find, by scale drawing, the total length of the horizontal and diagonal struts in one pylon. (Answer to the nearest foot.)

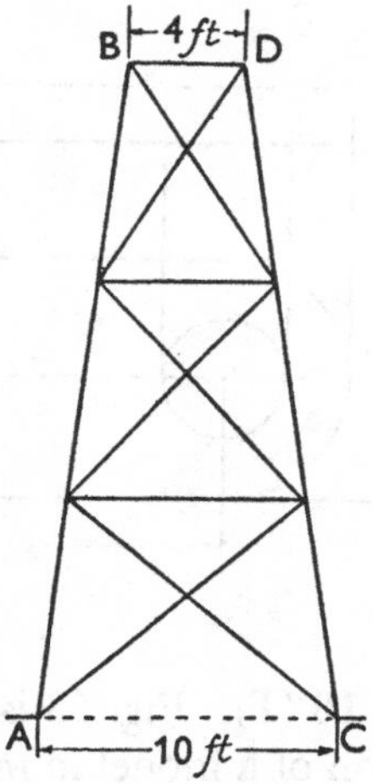

**8 (B).** The diagram (fig. 51) represents the plan and end elevation of the frame of a regular hipped roof. The wall plate is the top course of brickwork and *EC*, *BD* are the eaves.

*AB* and *AC* are the hip rafters and *KR* is the rise of the roof (i.e. height of ridge above the wall plate).

The length *OC* is called the 'eaves length' of the hip rafters and in a regular hipped roof this length is half the span (so that, in this case *OC* = *KG*).

Find, by drawing to scale, the actual length of a hip rafter in such a roof if the span is 30 ft. and the rise is 8 ft.

**9 (T).** Examine carefully the plan and elevations of a link plate as shown in fig. 52. The dimensions are shown on the drawing.

Fig. 50

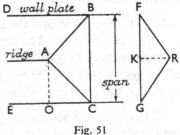

Fig. 51

What is (*a*) the overall length of the link plate, (*b*) the scale of the drawing? (*c*) Draw the plan to full size and sections through *AB* and *CD*.

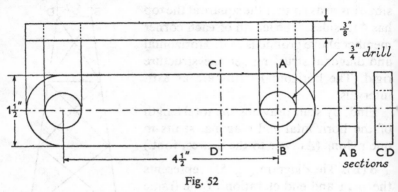

Fig. 52

**10 (T).** Fig. 53 is a scale drawing of the plan and side elevation of a model in metal. Some of the dimensions are not given.

    (*a*) State the scale of the drawings as a fraction of full size.

    (*b*) What is the true width of the top of the slot?

    (*c*) What is the true width of the bottom of the slot?

    (*d*) What is the true thickness of the model at *AB*?

    (*e*) Draw the plan and side elevation to full size and also a section through *AB*.

**11 (T).** Fig. 54 shows the plan and longitudinal elevation of a bicycle pedal-crank. The scale of the drawings is ½ full size.

Redraw these to ¾ full size and (*a*) state the full-size diameter of the gear-wheel spindle and also the pedal spindle; (*b*) how far apart are the centres of these two spindles?

**12 (B).** Fig. 55 shows the plan and elevation of a section of the corner of the wall plate fixed on the top course of brickwork

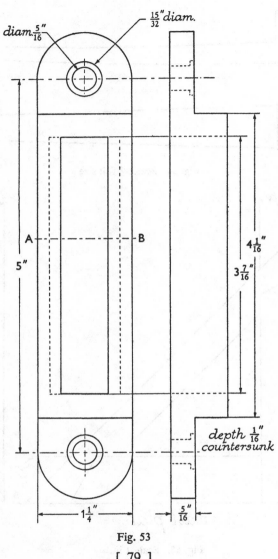

**Fig. 53**

[ 79 ]

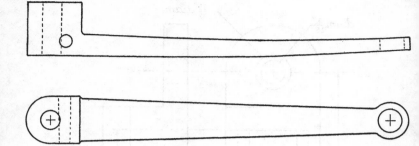

Fig. 54. Scale: ½ full size

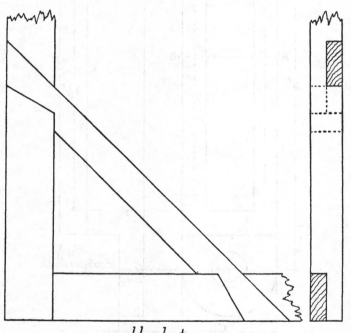

*wall plate*

Fig. 55. Scale 1 in. = 1 ft.

[ 80 ]

of an outside wall of a house to carry the rafters of the roof structure. The two timber members meet with a halving joint and the right angle is maintained by a dovetailed tie.

By measurement of the drawings determine the following: (*a*) what is the width and thickness of the timber used in the wall plate? (*b*) what is the width and thickness of the timber in the tie? Draw a plan, and front elevation of the tie to a scale of $1\frac{1}{2}$ in. = 1 ft.

**13** (T). The plan, elevation and section (fig. 56) illustrate a brass bush to be sunk as a bearing for a swivel pin.

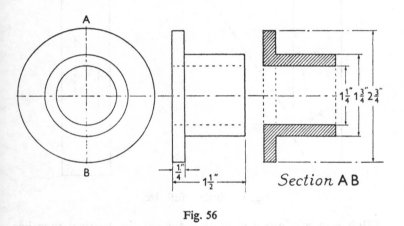

Fig. 56

The figures on the drawing are full-size dimensions.

Make a drawing for a similar bush $\frac{3}{4}$ full size and insert the corresponding dimensions.

As a check to your working, what are the following dimensions: (*a*) outside diameter of flange, (*b*) thickness of flange,

(c) size of drill necessary for the hole for bush to be inserted,
(d) depth to which bush is to be sunk?

**14** (B). The plan and elevation shown in fig. 57 are of
a wooden wall bracket to support a shelf.

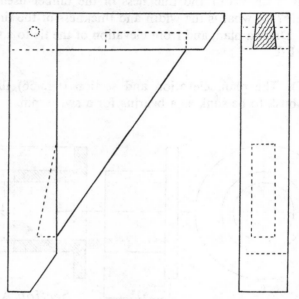

Fig. 57. Scale ¼ full size.

What size of timber is needed for (a) the vertical member,
(b) the horizontal member, (c) the support?

Draw plans and elevations of all three members separately,
ready for assembling, to a scale of ½ full size.

**15** (T). Fig. 58 shows the plan, side elevation and a section
of a bearing for a vertical spindle. The drawings are full size

but certain dimensions are missing. Obtain these by measurement and make drawings of a similar bearing 1¼ times as large. From measurement on your drawings state: (*a*) the overall thickness of the metal used for turning the bearing, (*b*) the depth to which the boring must be made to take the spindle.

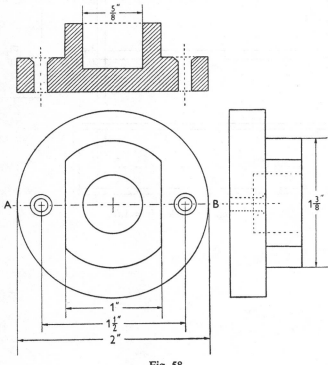

Fig. 58

**16** (B). From the drawings of a short, wooden form (fig. 59) estimate the shortest quantity of timber necessary if it is all supplied as 9 in. by 1 in. dressed.

Draw a plan, to a scale of $1\frac{1}{2}$ in. = 1 ft., of all the individual members before assembly.

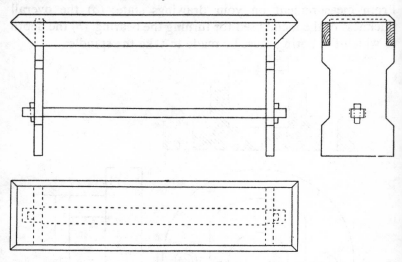

Fig. 59. Scale 1 in. = 1 ft.

CHAPTER 10

# MEASUREMENT OF AREAS AND VOLUMES

The areas of most plane figures as well as the volumes of regular or semi-regular solids are easily found from the measurement of certain dimensions and the use of simple formulae.

All that is necessary is to substitute these measurements in place of the letters used in the particular formula.

As a simple example, the area of a square or rectangular surface is found from the product of the length of two adjacent sides (base and height). If these two dimensions are denoted by $B$ and $H$ units of length respectively, then the area in each case is $B \times H$ *square* units of area.

In the case of a square we know that $B = H$ so the formula is $B \times B = B^2$ square units for the area of a square.

Particular care must be taken to see that the units of length, when substituting in a formula, are the same for all measurements taken, i.e. all in inches, so that the area is in square inches, or all in feet so that the area is in square feet, etc.

Many of the following formulae are simple enough to commit to memory, but others, which will be introduced later, need not be remembered—they can always be found when wanted in a book of tables or standard formulae.

## Areas of Plane Surfaces

The following are some of the more common plane surfaces that are likely to be met with in different branches of industrial work.

(i) *Parallelogram.* This is a plane four-sided figure with its opposite sides parallel but its angles not necessarily right angles (fig. 60).

Area $A = B \times H$ sq. units, where $H$ is the *perpendicular* height of the parallelogram.

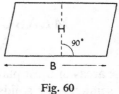

Fig. 60

(ii) *Triangle.* Any figure bounded by three straight lines is a triangle. If one of the angles is a right angle (90°) the triangle is a *right-angled triangle* (fig. 61). If one of the angles is greater than a right angle the triangle is an *obtuse-angled triangle*

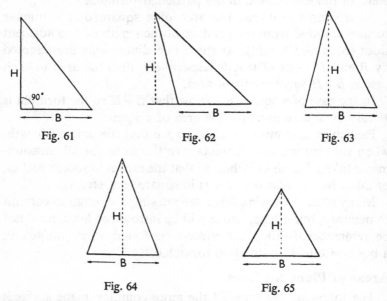

Fig. 61          Fig. 62          Fig. 63

Fig. 64          Fig. 65

(fig. 62). If all the angles are separately less than a right angle the triangle is an *acute-angled triangle* (fig. 63). An *isosceles*

[ 86 ]

*triangle* is one in which the angles at the base are equal (fig. 64). If all the angles are equal the triangle is an *equilateral triangle* (fig. 65).

The sum of all the angles of a triangle is always equal to *two right angles* (180°).

In every case the formula for the area is

$$A = \frac{B \times H}{2} \text{ sq. units,}$$

where $H$ is again the perpendicular height.

(iii) *Trapezoid.* This is a four-sided plane figure with one pair of opposite sides parallel (fig. 66).

In this figure:

$$A = \left(\frac{a+b}{2}\right) \times H \text{ sq. units.}$$

(iv) *Any irregular four-sided plane figure.* If we know the length of one *diagonal* ($L$) (fig. 67) and the lengths of the perpendiculars drawn upon it from the opposite angles, we have, in effect, two triangles, so that

$$A = \frac{a \times L}{2} + \frac{b \times L}{2} = \frac{L(a+b)}{2} \text{ sq. units.}$$

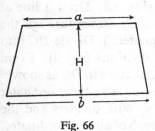

Fig. 66

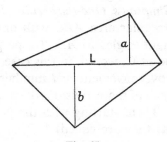

Fig. 67

(v) *Circle.* If the radius is known, then

$$A = \pi R \times R = \pi R^2 \text{ sq. units,}$$

where $\pi$ is the symbol denoting the number of times that the *diameter* divides into the *circumference*. It has the same value for all circles and is approximately 3·14 (fig. 68).

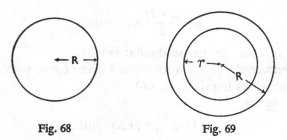

| Fig. 68 | Fig. 69 |

(vi) *The Plane Ring.* The area of the ring (fig. 69) is obviously the difference between the area of the outer circle and the area of the inner circle.

So that
$$A = \pi R^2 - \pi r^2$$

$$= \pi (R^2 - r^2) \text{ sq. units.}$$

(vii) *Any irregular area with one or more curved sides. Simpson's One-third rule.* Suppose that fig. 70 represents an irregular area *ABC* with one straight side *AB*. Draw a line at right angles to *AB*. (A perpendicular *DC*, drawn through the point *C*, would, in this case, simplify matters.) Divide *DC* into any convenient *odd* number of equal divisions (say 9). From each of these divisions draw a line perpendicular to *DC* as shown.

Then, starting with the point *D* as zero, label these *ordinates* (as they are called) 1, 2, 3, ..., up to 9. With dividers and the diagonal scale measure accurately the length of all these ordinates.

[ 88 ]

Add together the lengths of end ordinates nos. 0 and 9 (known as the *terminal* ordinates). (In this case the length of no. 9 is nil.) Call this total $S_T$.

Add together the sum of all the *even*-numbered ordinates (nos. 2, 4, 6 and 8). Call this total $S_E$.

Add together the sum of all the *odd*-numbered ordinates (nos. 1, 3, 5 and 7). Call this total $S_0$.

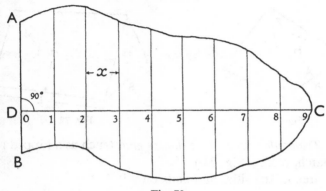

Fig. 70

Then the area of this irregular plane figure is found from the formula:

$$A = \frac{x}{3}(S_T + 2S_E + 4S_0) \text{ sq. units,}$$

where $x$ is the distance between any two consecutive ordinates.

## Areas of Surfaces of Solids

(i) *Square Pyramid.* Such figures are met with as caps topping a pillar, rick-thatching, etc. (fig. 71).

The total area of the four sloping sides is

$$A = 4\left(\frac{B \times H}{2}\right) = 2BH \text{ sq. units.}$$

[ 89 ]

If the base is included the area becomes

$$A = 2BH + B^2 = 2B(H + B) \text{ sq. units.}$$

(ii) *Triangular Prism* (as in any ordinary trussed roof, fig. 72). The area of the sides and ends only is

$$A = 2\left(\frac{B \times H}{2} + aL\right) = BH + 2aL \text{ sq. units.}$$

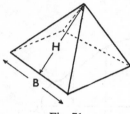

Fig. 71

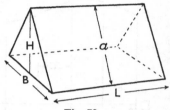

Fig. 72

(iii) *Triangular Prism with sloping ends* (such as a hipped roof, rick thatch, etc.). (Fig. 73.)

The area of the sides and ends only is

$$A = 2\left(\frac{BH}{2} + \frac{(R + L)a}{2}\right) = BH + a(R + L) \text{ sq. units,}$$

where $R$ is the length of the ridge.

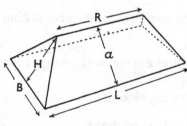

Fig. 73

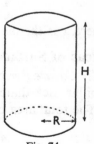

Fig. 74

[ 90 ]

# MEASUREMENT OF AREAS AND VOLUMES

(iv) *Cylinder* (such as roller bearings, cylindrical columns, crushing mills, etc.). (Fig. 74.)

The area of the curved surface only is

$$A = 2\pi RH \text{ sq. units.}$$

If the base and top are included, the total area is

$$A = 2\pi RH + 2\pi R^2$$

$$= 2\pi R (H + R) \text{ sq. units.}$$

(v) *Cone* (rick thatch of cylindrical stack, types of bearing, etc.). (Fig. 75.)

The area of the curved surface is

$$A = \pi RH \text{ sq. units.}$$

The area, including the base, is

$$A = \pi RH + \pi R^2$$

$$= \pi R (H + R) \text{ sq. units.}$$

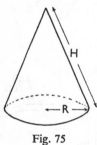

Fig. 75

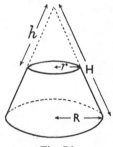

Fig. 76

(vi) *Truncated Cone*. This is a cone with a smaller cone removed from the top, as in a circular-rotating grinding mill, cotter spindle, etc. (Fig. 76.)

WHM

[ 91 ]

4

The area of the curved surface is obviously the difference between the areas of the curved surface of the large cone and the small cone.

So that $A = \pi RH - \pi rh$

$= \pi (RH - rh)$ sq. units.

(vii) *Sphere.* $A = 4\pi R^2$ sq. units.

# EXERCISE 11

Where a diagram is not given it is helpful to make a rough sketch to assist the easy working of the problem.

**1** (A). By measurement of the plan of a meadow (*A*) in fig. 77 determine the acreage (correct to the nearest first decimal place) of the area including the pond. (1 acre = 4840 sq.yd.)

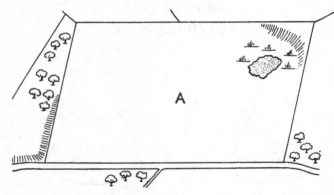

Fig. 77. Scale: 66 yd. = 1 in.

**2** (T). A length of circular shafting revolves in a brass bearing. The inside diameter of the bearing is $\frac{7}{8}$ in. and its length is $1\frac{1}{8}$ in. What is the area of rubbing surface to be lubricated? $(\pi = \frac{22}{7}.)$

**3 (A).** A hayrick, with a square base, measures 25 ft. along each of the four 'eaves'. The top of the rick is sloped to a point which is 12 ft. from the middle of the eaves on all sides. What area of thatching in 'squares' will be needed for the rick if a square of thatching is 100 sq.ft.?

**4 (T).** What is the area of one side of the link plate (fig. 78), if $\pi = 3 \cdot 14$? (Answer to nearest second decimal place.)

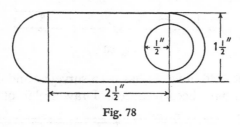

Fig. 78

**5 (B).** A hipped roof as shown in fig. 79 is to be tiled with tiles that cover six 'squares' per 1000 tiles. How many tiles will be needed?

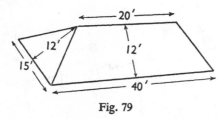

Fig. 79

**6 (T).** A hollow metal cone with open end, has a base diameter of 3 ft. 6 in. and a slant height of 5 ft. The thickness of the metal may be neglected. If this cone is to be chromium plated, inside and out, and the estimate for plating is 2s. per sq.ft., what would the job cost? ($\pi = \frac{22}{7}$.)

4-2

**7** (B). The diagram in fig. 80 represents the end wall of a building. What is its area? Draw this to a scale of 1 in. = 5 ft. and compare your answer with that obtained from working by Simpson's rule.

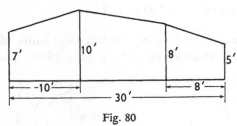

Fig. 80

**8** (A). The field in fig. 81 has been surveyed and the measurements shown have been recorded. Draw a plan of this field to

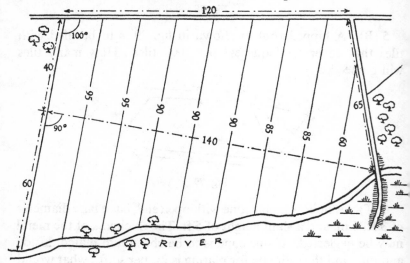

Fig. 81. Measurements in yards.

[ 94 ]

a scale of 1 in. = 20 yd. and estimate its acreage, to the nearest first decimal place, by means of Simpson's rule.

**9** (Tr). The silk canopy of a parachute when opened flat is seen to be circular in shape with a vent-hole in the centre. If the diameter of the parachute is 24 ft. and the diameter of the vent-hole is 2 ft., what is the area of silk, to the nearest square yard, used in its manufacture, if we allow an extra 5 sq.yd. for seams? ($\pi = 3 \cdot 14$.)

**10** (Tr). A motor-car tyre is marked $4 \cdot 50 \times 16$. This means that the tread of the tyre is $4 \cdot 5$ in. wide and the overall radius of the tyre is 16 in. When the tyre is properly inflated the tread thus forms what may be considered as the curved surface of a very short cylinder. Calculate the area of rubber subject to wear from road friction in this tyre. ($\pi = 3 \cdot 14$.)

**11** (B). A hall measures 12 ft. by 10 ft. 6 in. and is 10 ft. high from floor to ceiling. The lower edge of a cornice which surrounds the walls is 1 ft. 6 in. from the ceiling and the skirting is 6 in. from the floor. The door frame measures 6 ft. 6 in. by 3 ft. There is a fire grate 4 ft. wide and 3 ft. 6 in. high in one wall and in the opposite wall a window frame, which reaches to the ceiling, is 4 ft. wide and 7 ft. 6 in. high. The hall is to be panelled, between the cornice and skirting, in expensive timber. What area, to the nearest 10 square feet, has to be covered?

**12** (T). In electrical work the diameters of electrical leads are often expressed in *mils*, where 1 mil = 0·001 in. Their areas are then given in what are called 'circular mils', where a circular mil is the area of a circle of diameter 1 mil (0·001 in.).

We know that a wire of 1 mil diameter has an area

$$\frac{\pi}{4} (1)^2 = \frac{\pi}{4} \text{ sq.mils} = 1 \text{ circular mil.}$$

Therefore a wire of 2 mils diameter has an area of

$$\frac{\pi}{4}(2)^2 = 4\frac{\pi}{4}\text{ sq.mils} = 4\text{ circular mils}$$

and a wire of 3 mils diameter has an area of

$$\frac{\pi}{4}(3)^2 = 9\frac{\pi}{4}\text{ sq.mils} = 9\text{ circular mils}$$

and so on.

Thus the area in circular mils of any wire is the square of its diameter in mils.

(a) Find the cross-sectional area of a wire of 19 I.W.G. of diameter 0·04 in. (i) in square inches, (ii) in circular mils.

(b) A conductor is made of eight strands of 33 I.W.G., so that each strand has a diameter of 0·01 in. What is the total cross-sectional area of this conductor (i) in square inches, (ii) in circular mils. ($\pi = 3\cdot14$.) (Answers to three significant figures.)

## Volume

The volume of a solid is the amount of space that it occupies. In the same way the volume of a hollow container is the amount of space available to be filled, called its *capacity*. A knowledge of the volumes of some of the simple regular or semi-regular solids and containers is frequently required in industry and all may be obtained by using easy formulae.

To find the volume of any simple solid it is always necessary to have three measurements (or sometimes the same measurement more than once). As, in finding an area, we had two dimensions giving the answer in square units, so, in finding volume or capacity we have three dimensions and the answer is always in *cubic* units, e.g. cubic inches, cubic feet, etc. These, in the case of capacities, may be converted into the more familiar gallons, bushels, etc., if necessary.

As before, all measurements for the determination of a volume must be in the same units.

(i) *Rectangular block or tank*. (Fig. 82.)

$$V = L \times B \times H \text{ cu. units.}$$

We know that

$$L \times B = \text{area of base in sq. units.}$$

So that our formula becomes

$$V = \text{area of base} \times H.$$

This is true of *all* solids whose sides are vertical to the base, such as (ii), (iii) and (iv) below.

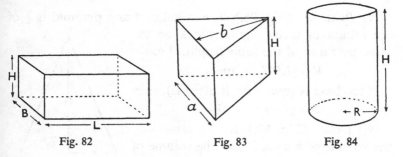

Fig. 82       Fig. 83       Fig. 84

(ii) *Triangular Prism*. (Fig. 83.)

$$V = \frac{ab}{2} \times H \text{ cu. units.}$$

(iii) *Cylinder*. (Fig. 84.)

$$V = \pi R^2 H \text{ cu. units.}$$

(iv) *Hollow Tubing*. A straight length of hollow tubing is, in effect, a cylinder with a smaller cylinder withdrawn from inside it (fig. 85). So that if the external radius is $R$ and the internal radius is $r$, then the volume, $V$, of the tubing itself is

$$V = \pi R^2 L - \pi r^2 L$$
$$= \pi L (R^2 - r^2)$$
$$= \pi L (R+r)(R-r) \text{ cu. units.}$$

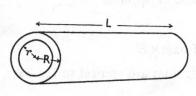

Fig. 85

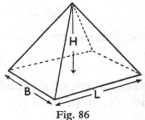

Fig. 86

(v) *Pyramid.* (Fig. 86.) The volume of any pyramid is $\frac{1}{3}$ of the volume of a rectangular block on the same base and of the same height. I.e.

$$V = \tfrac{1}{3} LBH \text{ cu. units.}$$

If the base is square (as it often is), then

$$V = \tfrac{1}{3} B^2 H \text{ cu. units.}$$

(vi) *Cone.* (Fig. 87.) In the same way the volume of a cone is $\frac{1}{3}$ of the volume of a cylinder on the same base and of the same vertical height. I.e.

$$V = \tfrac{1}{3} \pi R^2 H \text{ cu. units.}$$

Fig. 87

(vii) *Sphere.* In this case

$$V = \tfrac{4}{3} \pi R^3 \text{ cu. units.}$$

(viii) *Triangular Prism with sloping ends* (the hipped roof, fig. 88).

$$V = \frac{BH}{6}(2L + l) \text{ cu. units.}$$

This formula is perhaps best understood in this way. Suppose that we divide the solid into three parts by two cuts at right angles to the base, as shown in fig. 89.

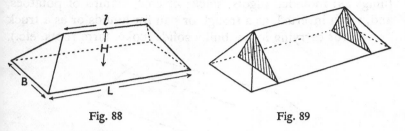

Fig. 88          Fig. 89

Place the two end pieces together and we have a prism and a pyramid (fig. 90), so that

$$V = \frac{BHl}{2} + \tfrac{1}{3}BH(L-l)$$

$$= \frac{BHl}{2} + \frac{BHL}{3} - \frac{BHl}{3}$$

$$= \frac{3BHl + 2BHL - 2BHl}{6}$$

$$= \frac{2BHL + BHl}{6} = \frac{BH}{6}(2L+l).$$

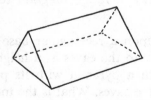
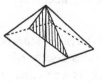

Fig. 90

(ix) *The Truncated Wedge.* This consists of a triangular sloping prism with its top removed by cutting parallel to the base, fig. 91. It is a shape very generally employed for such things as: moulded ingots, stacks of coal, clamps of potatoes, and, when inverted, as a trough or pan for liquids or as a truck or cart for carrying loose, bulky solids (coke, farm roots, etc.).

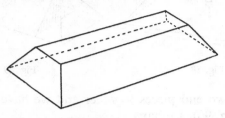

Fig. 91

The simplest formula for its volume is

$$V = \frac{H}{6}(A_1 + A_2 + 4M) \text{ cu. units,}$$

where $H$ is the vertical height, $A_1$ is the area of the base, $A_2$ is the area of the top, and $M$ is the area at the middle plane (i.e. where the vertical height is $H/2$).

## EXERCISE 12

Again, when necessary, draw a rough diagram to help the working.

1 (A). A hayrick stands upon a rectangular base 30 ft. by 21 ft. The sides rise vertically to the eaves at a height of 9 ft. and the rick is crowned with a pyramid with its point 9 ft. vertically above the level of the eaves. What is the total weight of the hay in the rick, allowing 10 cu.yd. to the ton?

**2 (B).** A hipped roof, with sloping ends, measures 36 ft. long by 20 ft. wide at the eaves. The ridge is 20 ft. long and is 12 ft. vertically above the level of the eaves. What is the storage space above the eaves?

**3 (T).** If cast iron weighs 456 lb. per cu.ft., find the weight of 1 yd. length of cast-iron piping of 6 in. internal bore and 1 in. thick. ($\pi = \frac{22}{7}$.)

**4 (Tr).** Tank lorries for distribution of motor spirit are often fitted with tanks which are oval in section. If the area of an oval (ellipse) is given by the formula $A = \pi ab$, where $a$ is half the major axis and $b$ is half the minor axis (fig. 92),

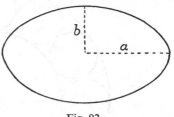

Fig. 92

find the capacity of such a tank 10 ft. 6 in. long with a maximum width of 7 ft. and a maximum depth of 3 ft. ($\pi = \frac{22}{7}$ and 1 cu.ft. = 6·25 gal.)

**5 (A).** A stable is rectangular and measures 30 ft. long by 20 ft. wide. The sides rise vertically to the eaves which are 9 ft. above ground-level. The roof is an ordinary ridged roof with the ridge 15 ft. from ground-level. The stabling accommodates five horses. Estimate (*a*) the floor area allowed for each horse, (*b*) the number of cubic feet of space allotted to each horse.

**6 (B).** An empty cylindrical hot-water tank and fittings weigh 47 lb. It measures 3 ft. 6 in. high and its internal diameter is 2 ft. 6 in. What weight must be allowed for when the tank is being fitted in position? (1 cu.ft. of water weighs $62\frac{1}{2}$ lb.; $\pi = \frac{22}{7}$.)

**7 (T).** Fig. 93 represents the section of a ball race with the dimensions shown. The housing for the balls is, in effect, a very

[ 101 ]

short length of hollow tubing. Calculate (*a*) the volume of grease needed for correct lubrication if the grease should fill completely the air spaces in the race. ($\pi = 3.14$.) (Answer to nearest cubic inch.) (*b*) What is the ratio of total volume of the balls to the volume of the 'housing' space?

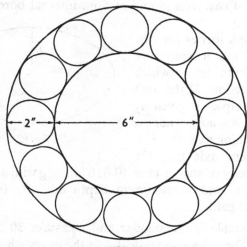

Fig. 93

**8** (Tr). A stack of engine coal measures at the oblong base 30 ft. by 21 ft. wide. The sides and ends slope regularly to the flat top face of the stack which is an oblong 20 ft. long by 14 ft. wide. The vertical height of the stack is 6 ft. Assuming that one ton of coal occupies a volume of 40 cu.ft., estimate the weight of coal in the stack.

**9** (B). The weight of any building structure is borne by the soil beneath the foundations. There is a limit to the pressure that any soil can withstand and to lessen the intensity of pressure the footings of a building are often splayed as shown in fig. 94. This diagram represents a granite pillar 20 ft. high and the footing is granite in the shape of a truncated square pyramid. If granite weighs 166 lb. per cu.ft., estimate (*a*) the weight in tons of the pillar and footing, (*b*) the pressure on the soil in tons per sq.ft.

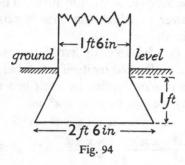

Fig. 94

**10** (T). What is the acid content in pints of an accumulator with internal measurement 10 in. by 7 in. by 16 in. deep, with nine plates 12 in. by 8 in. by ¼ in.? These are fixed so that they rest 1 in. from the bottom of the trough. The acid must be level with the top of the plates. (1 cu.ft. = 6¼ gal.)

## CHAPTER 11

# LOGARITHMS

When we wish to write a product such as $x \times x \times x \times x \times x$ we write it usually in the form $x^5$.

In this case we say that the *base* $x$ has been raised to the *power* 5, or more simply, $x$ 'to the fifth'. The small numeral indicating the power to which the base is raised is called the *index* (plural indices).

In passing, it is advisable to avoid the common habit of explaining $x^5$ as '$x$ multiplied by itself five times'. This is untrue. (How many times is $x$ multiplied by itself in $x \times x = x^2$?)

In the same way we can write $x^3 \times x^2$ as

$$x \times x \times x \times x \times x = x^5 = x^{3+2}$$

and
$$\frac{x^5}{x^2} = \frac{x \times x \times x \times x \cdot \not{x} \times \not{x}}{\not{x} \times \not{x}} = x^3 = x^{5-2}.$$

So that, when multiplying powers of the same base, we *add* the indices to obtain the final power; and when dividing we *subtract* the index of the denominator from that of the numerator.

These same rules apply when the base is a *numeral* instead of a symbol.

Thus
$$100 = 10 \times 10 = 10^2,$$
$$10,000 = 10 \times 10 \times 10 \times 10 = 10^4.$$

Hence
$$100 \times 10,000 = 10^2 \times 10^4 = 10^6,$$

and
$$\frac{10,000}{100} = \frac{10^4}{10^2} = 10^{4-2} = 10^2.$$

Now, if $10^2 = 100$, and $10^3 = 1000$, what is the value of $10^{2.5}$? Obviously it must be a number somewhere between 100 and 1000. Actually, by mathematical calculation the value of $10^{2.5}$ is found to be 352.

[ 104 ]

In the same way we could find the answer for any decimal value of the index.

For example, we could prove that

$$10^{2 \cdot 7259} \text{ is very nearly equal to 532,}$$

and $\qquad 10^{2 \cdot 9703} \qquad , , \qquad\qquad , , \qquad 934.$

(The index in these cases is only a close approximation since the decimal part is correct to four places only.)

So that any number, therefore, may be expressed as a base 10 raised to a suitable power.

When a number is expressed as a base 10 raised to any power, the index is called a *logarithm*. Thus in the expression $10^{2 \cdot 7259} = 532$, 2·7259 is called the logarithm of 532 to the base 10.

As all logarithms that we shall use are to the base 10, we omit reference to the base, and call them, simply, logarithms of the numbers. Therefore 2·7259 is the logarithm of 532 and the equation is written: log 532 = 2·7259, which means that the logarithm of 532 is 2·7259.

From the above we see that a logarithm can consist of two parts: (*a*) the *decimal* part, (*b*) the whole number, or *integral* part.

The decimal part of a logarithm is found from a table of logarithms, as supplied at the end of the book, where the logarithm of any number of not more than four digits (i.e. from 0 to 9999) can be found.

The integral part is determined by the use of two very simple rules which are best explained by examples.

*Example* 1. Suppose that we wish to find the logarithm of 2146.

First divide the number, mentally, into hundreds, tens and units, thus: 21/4/6. From the tables we then determine the decimal part of the logarithm in the following way.

Looking down the extreme left-hand column of the tables we find the number 21. Travelling horizontally along this line until we reach the vertical column '4', we read 3304. For the final digit 6, we continue along the horizontal '21' line until we come to the vertical column '6' in the 'differences'. Here we read 12. We notice also as we proceed across the horizontal line that the numbers increase, so that the difference 12 must be added to 3304 to give the final figure $3304 + 12 = 3316$. This is the decimal part of log 2146.

To obtain the integral part of the logarithm we apply these simple rules:

*Rule* 1. Count the number of digits in the whole number part of the quantity in question and subtract 1. In this case 2146 is a whole number of four digits. So the integral part of the logarithm is $4 - 1 = 3$, and our final answer is

$$\log 2146 = 3 \cdot 3316.$$

Suppose now that we want to find log 21·46. Taking no notice of the decimal point we read from the tables the *same* value for the decimal part of the logarithm, and in this case, as there are two digits in the whole number part of 21·46, the logarithm has $2 - 1 = 1$ as the integral portion. So that $\log 21 \cdot 46 = 1 \cdot 3316$.

*Example* 2. To obtain the integral part of the logarithm of a decimal with *no* whole number in front of the decimal point. We have just shown that $\log 21 \cdot 46 = 1 \cdot 3316$ (with $2 - 1 = 1$ as the integral part).

Similarly, $\log 2 \cdot 146 = 0 \cdot 3316$ (with $1 - 1 = 0$ as the integral part).

In the same way the integral part of log 0·2146 is $0 - 1 = -1$. (This is always written $\bar{1}$ to show that the integral part *only* is negative. The decimal part of a logarithm is *always* positive.) So that $\log 0 \cdot 2146 = \bar{1} \cdot 3316$.

We see from the above that every time we divide the given number by 10 we reduce the integral part of the logarithm by 1. If we divide 0·2146 by 10 we get 0·02146 and the whole number part of the logarithm becomes $\bar{1} - 1 = \bar{2}$.

For 0·002146 the logarithm has an integral part $\bar{3}$, and so on.

*Rule* 2, therefore, is as follows: To find the integral part of the logarithm of a decimal, count the number of 0's between the decimal point and the first significant figure, and *add* 1. This figure so obtained is always *negative* or minus.

## Multiplication and Division by Logarithms

It is as a 'short-cut' method of multiplying and dividing that we shall find logarithms of most use to us.

*Example* 3. Suppose that we wish to multiply 532 by 934.

We know that $532 = 10^{2·7259}$ and $934 = 10^{2·9703}$. Therefore

$$532 \times 934 = 10^{2·7259} \times 10^{2·9703}$$
$$= 10^{5·6962}.$$

Until quite familiar with the use of logarithms the student is recommended to work with powers of 10. Later on, after sufficient practice to give confidence, the above may be written as:

$$\log (532 \times 934) = \log 532 + \log 934$$
$$= 2·7259 + 2·9703$$
$$= 5·6962.$$

The danger to a beginner in writing the problem in logarithm form is that he may, and often does, leave the answer as 5·6962. This is still more likely to be done when multiplying or dividing decimals.

What we have actually found is the logarithm of the answer, so we must refer again to the tables to find the number whose logarithm is 5·6962.

This time we work backwards and, taking notice of the decimal part only, we find the nearest number below this in the *body* of the tables. This is 6955 in the horizontal column 49 and in the vertical column 6. The difference between 6955 and 6962 is 7 and in the 'differences' (still keeping along the same horizontal line) we find 8 at the top of the corresponding vertical column. So that the figures in the answer are 4968.

By rule 1, we know that the integral part of the logarithm, 5, is one *less* than the number of digits in the answer, which must therefore have *six* digits, and the product $532 \times 934 = 496,800$ is the required answer.

By actual multiplication we find the correct answer is 496,888, so that the answers obtained by four-figure logarithms are in nearly all cases only close approximations.

*Example* 4. What is the value of

$$\frac{23 \cdot 62 \times 0 \cdot 1412}{162 \cdot 5}?$$

By reference to logarithm tables this may be written

$$\frac{10^{1 \cdot 3733} \times 10^{\bar{1} \cdot 1498}}{10^{2 \cdot 2108}} = 10^{(1 \cdot 3733 + \bar{1} \cdot 1498 - 2 \cdot 2108)}. \tag{i}$$

Here is an example that contains integral parts of the logarithms with both $+$ and $-$ sign. To avoid error it is advisable to work these independently in the margin.

Thus, collecting the whole numbers, we have

$$1 + \bar{1} - 2 = 1 - 1 - 2 = 1 - 3 = \bar{2}.$$

Simplifying the decimal part, we have

$$0 \cdot 3733 + 0 \cdot 1498 - 0 \cdot 2108 = 0 \cdot 5231 - 0 \cdot 2108$$
$$= 0 \cdot 3123.$$

Therefore the expression (i) becomes

$$10^{\bar{2}\cdot3123}.$$

Again by reference to the tables and rule 2 we find the number whose logarithm is $\bar{2}\cdot3123$ is $0\cdot02052$.

Considerable care must always be taken in the addition and subtraction of logarithms. For example, what is the value of

$$0\cdot2162-\bar{3}\cdot8273+2\cdot1614-1\cdot0021?$$

Again, taking the integral parts first, we have:

$$0-\bar{3}+2-1=0+3+2-1=5-1=4.$$

Simplifying the decimal parts, we have:

$$0\cdot2162-0\cdot8273+0\cdot1614-0\cdot0021$$
$$=0\cdot3776-0\cdot8294.$$

Here we must remember that the fractional part of any logarithm *must* be positive (+) so that we must not subtract $0\cdot3776$ from $0\cdot8294$ as the remainder would be negative (−). Therefore we 'borrow' one from the integral part, 4, leaving 3 and obtain

$$1\cdot3776-0\cdot8294=0\cdot5482,$$

and the final value of the logarithm is $3\cdot5482$. Also, remember that if we 'borrow' one from a negative integer, say $\bar{3}$, we leave $\bar{4}$ as the final integer.

For example, what is the value of

$$\bar{3}\cdot2136-1\cdot5224?$$
$$\bar{3}\cdot2136-1\cdot5224=(-3-1)+0\cdot2136-0\cdot5224$$
$$=\bar{4}+0\cdot2136-0\cdot5224$$
$$=\bar{5}+1\cdot2136-0\cdot5224$$
$$=\bar{5}\cdot6912.$$

[ 109 ]

Lastly, what is the value of log 1?

We know that we can multiply any number or decimal by 1 without altering its value.

For example, $2 \cdot 35 = 2 \cdot 35 \times 1$.

Now if we take logarithms, we have

$$\log 2 \cdot 35 = \log (2 \cdot 35 \times 1)$$
$$= \log 2 \cdot 35 + \log 1,$$

and the only way that this equation can be true is for log 1 to be equal to 0.

Therefore $\log 1 = 0$, or in other words $10^0 = 1$.

## EXERCISE 13
(Assume, when necessary, that $\pi = 3 \cdot 142$.)

**1.** Use the logarithm tables to check the following:

| Number | Decimal for first three figures | Added difference | Full decimal | Integral part |
|--------|---------------------------------|------------------|--------------|---------------|
| 1122 | 0·0492 | 8 | 0·0500 | 3 |
| 1554 | 0·1903 | 11 | 0·1914 | 3 |
| 30·3 | 0·4814 | 0 | 0·4814 | 1 |
| 5·678 | 0·7536 | 6 | 0·7542 | 0 |

**2.** From the tables write down the logarithms of the following numbers: (a) 3468, (b) 346·8, (c) 3·468, (d) 76·42, (e) 7·642, (f) 81·79, (g) 220·7, (h) 2200, (i) 23, (j) 3, (k) 200, (l) 4001, (m) 400·1, (n) 73250, (o) 897000, (p) 47287.

**3.** From the tables write down the numbers whose logarithms are the following: (a) 1·8982, (b) $\overline{1}$·5069, (c) 0·9085, (d) $\overline{2}$·8904, (e) $\overline{3}$·4771, (f) 2·9150, (g) $\overline{4}$·9031, (h) $\overline{1}$·6128, (i) $\overline{2}$·3222, (j) $\overline{3}$·4801, (k) $\overline{1}$·7243.

**4.** Find the answers to the following products by logarithms:

(a) $54 \cdot 41 \times 37 \cdot 13$,

(b) $2 \cdot 3 \times 4 \cdot 92 \times 31 \cdot 63 \times 0 \cdot 009$,

(c) $45 \cdot 2 \times 0 \cdot 913 \times 0 \cdot 05 \times 1 \cdot 212$,

(d) $0 \cdot 007 \times 0 \cdot 02 \times 0 \cdot 006$,

(e) $23 \cdot 74 \times 0 \cdot 0005 \times 2 \cdot 621 \times 3 \cdot 753$.

**5.** Use logarithms to simplify the following expressions:

(a) $\dfrac{31 \cdot 78}{6 \cdot 245}$,     (b) $\dfrac{2041}{13 \cdot 5}$,     (c) $\dfrac{2 \cdot 4}{0 \cdot 0072}$,

(d) $\dfrac{0 \cdot 014}{31 \cdot 23}$,     (e) $\dfrac{0 \cdot 061}{0 \cdot 0824}$,     (f) $\dfrac{0 \cdot 3714}{0 \cdot 1212}$,

(g) $\dfrac{29 \cdot 21 \times 11 \cdot 6}{102 \cdot 4}$,     (h) $\dfrac{1 \cdot 33 \times 11 \cdot 28}{0 \cdot 4}$,

(i) $\dfrac{112 \cdot 8 \times 0 \cdot 45}{1 \cdot 133}$,     (j) $\dfrac{209 \times 3 \cdot 26 \times 0 \cdot 08}{4 \cdot 7 \times 8 \cdot 213}$,

(k) $\dfrac{20 \cdot 3 \times 1 \cdot 781}{6 \cdot 235 \times 0 \cdot 9 \times 0 \cdot 062}$,     (l) $\dfrac{1}{5 \cdot 871 \times 1 \cdot 66}$.

**6 (Tr).** What is the circumference of a motor tyre whose radius is 16 in.?

**7 (B).** A rectangular tank is 3 ft. 6 in. long by 2 ft. 3 in. wide and 1 ft. 8 in. deep. What is its capacity in gallons if 1 cu.ft. = $6 \cdot 25$ gal.?

**8 (T).** No. 34 s.w.g. copper wire offers an electrical resistance of $636 \cdot 6$ ohms per mile length. What is the resistance of 1000 yd.?

**9 (T).** A small casting of aluminium alloy measuring $12 \cdot 6$ cu.in. weighs $1 \cdot 197$ lb. Find the weight per cu.ft. of this alloy.

**10** (B). A sheet of lead of thickness 0·16 in. measures 8 ft. by 6 ft. and weighs 457·9 lb. What is its weight per sq.yd.?

**11** (T). No. 15 S.W.G. copper wire of cross-section 0·00408 sq.in. carries with safety a maximum current of 23·2 amp. What is the equivalent maximum current in amp. per sq.in.?

**12** (B). A garage 22 ft. 6 in. by 12 ft. 9 in. is to be floored with concrete 4 in. thick. How many cubic yards of wet mixed concrete are needed? (Answer to nearest ½ yd. permissible.)

**13** (A). A rain-gauge during one night records a rainfall of 0·5 in. What does this represent in tons per acre? (1 cu.ft. of water weighs 62·5 lb.)

**14** (T). A rectangular copper tank, open at the top, measures 3 ft. 9 in. long, by 3 ft. 3 in. wide and 2 ft. 9 in. deep. It is constructed of sheet copper weighing 12 lb. per sq.ft. Find (*a*) the surface area of the tank in square feet, (*b*) the weight of the tank in cwt. allowing 2 % extra for lapped seams, (*c*) its gross weight when filled with water.

**15** (A). A farmer manured a field with a dressing of artificial manure at the rate of 1½ cwt. of nitrate of soda, 2 cwt. of superphosphate and 3 cwt. of Kainit per acre. What is the approximate distribution in oz. per sq.yd.?

**16** (B). A square column of granite is 10 ft. high and each side of the base measures 1 ft. 6 in. Find the pressure on the ground, in lb. per sq.ft., (*a*) when upright, (*b*) when lying flat. (Granite weighs 166 lb. per cu.ft.)

**17** (T). The electrical resistance $R$ (ohms) of a conducting wire is obtained from a formula

$$R = \frac{L \times P}{A},$$

where $L$ is the length of the wire in inches; $A$ is the cross-section of the wire in square inches; $P$ is a number depending upon the metal of which the wire is manufactured. What is the resistance of 1000 yd. of copper wire of cross-section 0·005 sq.in. when $P = 0·00000066$?

**18** (B). To determine $S$, the number of standard rods of brickwork in a wall 13½ in. (one brick and a half) thick, the following formula may be used:

$$S = L \times H \times 0·00368,$$

where $L$ is the length and $H$ the height of the wall, both measured in feet.

How many rods of brickwork are there in a 13½ in. wall which measures 16 ft. 9 in. by 8 ft. 3 in.?

**19** (A). A corn-rick, built on a circular base, has a diameter of 21 ft. at the eaves and is capped by a cone of slant height 12 ft. What is the area to be thatched to the nearest square yard?

**20** (T). The safe working load of a wrought-iron chain is 0·167 of its breaking strain. If the breaking strain is 32·35 tons, what is its safe working load?

## Square and Square Root

We know that when a number is multiplied by itself it is said to be squared. Suppose that the number is $x$. Then $x \times x = x^2$.

Conversely, if we wish to find a number which when multiplied by itself will equal a given number, then we find what is called the *square root* of the given number.

Suppose again that the given number is $x$. Then we know that the only number which when multiplied by itself equals $x$, must be $x^{\frac{1}{2}}$, since $x^{\frac{1}{2}} \times x^{\frac{1}{2}} = x^{\frac{1}{2}+\frac{1}{2}} = x$. Therefore the square root of any number is that same number raised to the power $\frac{1}{2}$.

Many engineering and similar problems call for the determination of square roots and although these can be calculated arithmetically or from specially prepared tables of square roots, it is simpler and quicker to obtain answers by the use of logarithms.

We know that the logarithm of the product of two or more numbers is the sum of their logarithms. Suppose that two numbers in a product are the same, namely, $x^{\frac{1}{2}}$ (when $x$ is any number). So that

$$x^{\frac{1}{2}} \times x^{\frac{1}{2}} = x.$$

Then $$\log x^{\frac{1}{2}} + \log x^{\frac{1}{2}} = \log x$$

and $$2 \log x^{\frac{1}{2}} = \log x.$$

Therefore $$\log x^{\frac{1}{2}} = \tfrac{1}{2} \log x,$$

and we consequently have a simple division by 2 of the logarithm of the given number to give us the logarithm of its square root.

*Example*. What is the square root of 23·56? (Usually written $\sqrt{23·56}$.) In other words what is the value of $23·56^{\frac{1}{2}}$.

By logarithms we know that

$$\log 23·56^{\frac{1}{2}} = \tfrac{1}{2} \log 23·56$$
$$= \tfrac{1}{2} \times 1·3722$$
$$= 0·6861$$

and the number whose logarithm is 0·6861 is (from the tables) 4·854.

In the same way the logarithm of the *cube* root of a number, such as $x^{\frac{1}{3}}$, is $\tfrac{1}{3} \log x$.

[ 114 ]

# LOGARITHMS

*Example* 1. The cross-sectional area of a heavy electrical lead is known to be 0·35 sq.in. What is its diameter to the nearest thousandth of an inch?

We know from the formula, that area $= \pi r^2$.

Therefore $\pi r^2 = 0 \cdot 35$ sq.in. and $r^2 = \dfrac{0 \cdot 35}{3 \cdot 142}$ sq.in.

Taking logarithms, we have

$$\log r^2 = \log 0 \cdot 35 - \log 3 \cdot 142.$$

I.e. $\qquad 2 \log r = \bar{1} \cdot 5441 - 0 \cdot 4972$

$$= \bar{1} \cdot 0469$$

and $\qquad\qquad \log r = \dfrac{\bar{1} \cdot 0469}{2}.$

In this case direct division of $\bar{1} \cdot 0469$ by 2 would leave $\bar{\tfrac{1}{2}}$ as the integral part of the logarithm, so we must rewrite the numerator in this form: $\bar{1} - 1 + 1 \cdot 0469 = \bar{2} + 1 \cdot 0469$. This has not altered its value (since $-1 + 1 = 0$) and we have

$$\log r = \frac{\bar{2} + 1 \cdot 0469}{2}$$

$$= \bar{1} \cdot 5235.$$

Therefore $r = 0 \cdot 3339$ in. (from the tables). The diameter of the lead is therefore

$2 \times 0 \cdot 3339$ in. $= 0 \cdot 6678$ in.

$\qquad\qquad = 0 \cdot 668$ in.   (to the nearest thousandth).

## EXERCISE 14

(Use $\pi = 3\cdot142$ when necessary.)

1. Find the value of

(a) $7\cdot532^2$,  (b) $0\cdot218^2$,  (c) $0\cdot316^3$,

(d) $\sqrt{109\cdot7}$,  (e) $4\cdot55^2 \times \sqrt{16\cdot52}$,  (f) $\dfrac{1\cdot835^3}{\sqrt{21\cdot9}}$.

2 (A). A horse is tethered to a peg on an open common by a rope 15 ft. in length. What is its approximate grazing area in square yards?

3 (T). A circular disc of metal of diameter 1·75 in. is to be drilled to make a washer so that the area of the hole to be drilled is half the area of the disc. What is the diameter of the hole?

4 (T). An ordinary electric light bulb is so designed that it takes sufficient current from the supply mains to raise the temperature of the metallic filament to glowing white heat (incandescence). The electrical resistance of the filament when hot is much greater than when cold. The power needed to bring the lamp to incandescence is measured in *watts* and is marked on the bulb of the lamp (40 W., 60 W., etc.). For any incandescent electric lamp the following relation holds good between the current in amperes ($I$), the hot resistance of the lamp in ohms ($R$) and the consumption of the lamp in watts ($W$):

$$I = \sqrt{\frac{W}{R}}.$$

Find the current consumption of (a) an 80-watt lamp with hot resistance 605 ohms, (b) a 150-watt lamp with hot resistance 323 ohms.

**5 (B).** A hemispherical dome, measuring 30 ft. in diameter, is to be covered with lead sheeting weighing 6 lb. per sq.ft. What weight of lead, to the nearest cwt., will be required for the job?

**6 (T).** In order to determine the correct size of rivets for riveting boiler plates so that the joints shall have sufficient strength, the following formula is employed:

$$\text{Diameter of rivet} = 1 \cdot 2 \sqrt{t},$$

where $t$ is the thickness of the boiler plate.

What size rivets should be used for boiler plates $\frac{3}{4}$ in. thick?

**7 (Tr).** The distance in which a car can be brought to a stop, without skidding, by the application of brakes, depends upon the speed of the car. It depends also upon the weight of the car, condition of the road surface, type of tyre and other considerations. Let us suppose that for a certain type of car, under good conditions:

$$D = \frac{V^2}{56},$$

where $D$ is the distance in yards that the car covers after the brakes have been fully applied and $V$ is the speed of the car in m.p.h.

What is the minimum distance in which this car could be pulled up when travelling at 42 m.p.h.?

**8 (T).** Television waves travel in straight lines but, unlike ordinary broadcast sound waves, they are not reflected during their progress and, in consequence, reception can be certain only at places which are not 'out of range' by reason of the curvature of the earth. Naturally, the higher the sending aerial and the higher the receiving aerial, the greater the range of reception.

The distance of reception, in miles, at ground-level, may be roughly estimated from this simple formula:

$$D = 1.5 \sqrt{S},$$

where $S$ is the height of the sending aerial in feet.

Up to what distance would you expect to get good reception from a sending aerial at a height of (a) 775 ft., (b) 1023 ft.? (Answer to nearest mile.)

**9** (A). A flock of 148 sheep is to be contained in a square fold and each sheep is to be allowed 35 sq.ft. of feeding space. How many 6 ft. hurdles will be needed to build this fold?

**10** (T). For reasons of safety the maximum current that may be carried in a copper conductor without excessive heating is based upon a standard of 1000 amp. per sq.in. of cross-sectional area. The smaller the cross-sectional area the less the corresponding load.

What is the diameter of a copper wire that can safely carry a maximum current of 20 amp.? (Answer to nearest second decimal place.)

**11** (T). A casting of a bronze sphere is thought to be faulty and to contain a 'blow-hole' inside. If its diameter is 20 in. and the bronze is known to weigh 420 lb. per cu.ft., estimate the volume of air space if the casting weighs $7\frac{1}{2}$ cwt.

**12** (A). A rectangular grazing area measuring 210 yd. by 75 yd. is to be fenced in. What saving, in yards, in fencing would be made by enclosing the same area in the form of a square?

# CHAPTER 12

# FORMULAE

## Evaluation and Transposition

In the last few chapters the solution of problems involving areas, volumes, square roots and other simple applications has been effected by the use of easy formulae. So far, in all cases we have been able to make use of the formulae by direct substitution of numbers in place of the letters or symbols of the formula.

This substitution, or evaluation, is always necessary when applying a formula to a particular problem. What also has to be done very frequently is to *transpose* the formula so that the unknown factor is isolated and can then be more readily determined.

In the transposition of formulae a few elementary rules must be observed and these are best explained by a series of easy examples.

*Example* 1. Formulae involving multiplication or division only.

Let us examine the simple relationship $C = \dfrac{V}{R}$, which is a short way of expressing what is called Ohm's Law. This law states that the current flowing in a circuit measured in amperes ($C$) is always equal to the electrical 'pressure' measured in volts ($V$), divided by the resistance measured in ohms ($R$).

As it stands, the expression is most convenient for finding $C$ when $V$ and $R$ are both known. ($C$ is said to be the subject of the equation.)

Suppose that we know $C$ and $R$ and wish to find $V$.

In any such expression as this we may multiply or divide both sides by the same symbol or number without altering the equation. If we multiply both sides by $R$ we obtain

$$CR = \frac{\cancel{R}V}{\cancel{R}} = V,$$

i.e. $$V = CR,$$

so that we have now made $V$ the subject.

Similarly, to find $R$ when $C$ and $V$ are known we have

$$V = CR \quad \text{or} \quad CR = V.$$

Dividing both sides by $C$ we get

$$R = \frac{V}{C}.$$

*Example* 2. Formulae involving addition and subtraction.

The formula for use with a certain type of pulley and tackle is

$$P = \frac{W}{3} + 2,$$

where $P$ is the effort in lb., required to lift a weight of $W$ lb.

To find $W$ (when $P$ is known) first subtract 2 from both sides of the equation, i.e.

$$P - 2 = \frac{W}{3}.$$

Multiplying both sides by 3 we have

$$W = 3\,(P-2)$$
$$= 3P - 6.$$

*Rule*. When any 'term' (which means a symbol, or numeral or collection of both) is transferred from one side of an equation to the other its *sign* is changed.

*Example* 3. Formulae involving brackets.

From the formula $R_t = R_0 (1 + at)$, where $R_t$, the electrical resistance of a conductor at $t°$ C., is found from $R_0$, the electrical resistance at $0°$ C., $t$ the temperature in $°$ C. and $a$ the temperature coefficient of the metal, suppose that we wish to find $t$. In this case first treat the expression in the bracket independently and isolate this by dividing both sides by $R_0$, so that

$$(1 + at) = \frac{R_t}{R_0},$$

i.e.

$$at = \frac{R_t}{R_0} - 1.$$

And finally, dividing both sides by $a$, we get

$$t = \frac{1}{a}\left(\frac{R_t}{R_0} - 1\right).$$

This may, if desired, be written

$$t = \frac{R_t - R_0}{aR_0}.$$

*Example* 4. Formulae involving squares, square roots or higher powers.

The formula $A = \pi (R^2 - r^2)$ has already been used to find the cross-sectional area of a plane ring or pipe.

Let us suppose that $A$ and $R$ are known and $r$ (the internal radius of the pipe) is required.

First isolate the bracket by dividing both sides by $\pi$.

Then

$$(R^2 - r^2) = \frac{A}{\pi},$$

and

$$-r^2 = \frac{A}{\pi} - R^2.$$

[ 121 ]

By changing signs throughout, we have

$$r^2 = R^2 - \frac{A}{\pi}$$

$$= \frac{\pi R^2 - A}{\pi}.$$

Taking square roots,

$$r = \left(\frac{\pi R^2 - A}{\pi}\right)^{\frac{1}{2}},$$

which is a suitable expression, with the help of logarithm tables, for finding $r$.

In the case of a square-root formula, such as $d = 1 \cdot 2 \sqrt{t}$, when it is necessary to find $t$, first, as before, isolate the term $\sqrt{t}$ by dividing throughout by $1 \cdot 2$. So that

$$\sqrt{t} = \frac{d}{1 \cdot 2}.$$

Square both sides, to obtain $t$, and we have

$$t = \frac{d^2}{1 \cdot 2^2} = \frac{d^2}{1 \cdot 44}.$$

(Be careful when squaring a fraction not to omit to square the denominator as well as the numerator.)

*Example* 5. Formulae involving addition and subtraction of fractions.

Suppose that we wish to find $v$, from the equation

$$\frac{2}{r} = \frac{1}{v} + \frac{1}{u}.$$

In this case, transfer $\frac{1}{u}$ to the opposite side of the equation, so that

$$\frac{1}{v} = \frac{2}{r} - \frac{1}{u}.$$

[ 122 ]

Now simplify the expression $\dfrac{2}{r} - \dfrac{1}{u}$ by bringing it to a common denominator, thus

$$\frac{1}{v} = \frac{2u - r}{ru}.$$

Inverting both expressions

$$v = \frac{ru}{2u - r}.$$

*Important.* Be particularly careful to remember that the reciprocal of such an expression as

$$\frac{2}{r} - \frac{1}{u} \text{ is not } \frac{r}{2} - \frac{u}{1}.$$

Such expressions must always be collected into one single term before inverting.

## EXERCISE 15

In questions 1–4, transpose the formulae to isolate the required symbol, and in the remaining questions use logarithms when necessary.

1. (a) $C = 2\pi r$, find $r$;  (b) $s = u + ft$, find $f$;

   (c) $V = v(1 + at)$, find $a$;  (d) $C = \dfrac{nE}{nr + R}$, find $n$;

   (e) $S = \dfrac{W(v - u)}{2g}$, find $u$;  (f) $f^2 = \dfrac{1}{4\pi^2 LC}$, find $L$.

2. (a) $\dfrac{m}{n} = \dfrac{ax + b}{cx + d}$, find $x$;  (b) $c = \dfrac{ax - 2by}{ax + 2by}$, find $x$;

   (c) $V = \pi h^2 \left(r - \dfrac{h}{3}\right)$, find $r$;

   (d) $\dfrac{1}{p^2} = \dfrac{1}{a} + \dfrac{1}{b}$, find $p$, $a$ and $b$.

**3.** If $\dfrac{2}{R} = \dfrac{1}{a} + \dfrac{1}{b}$, and $m = \dfrac{a-R}{R-b}$, find a value for $m$ which does not contain $R$.

**4.** If $s = ut + \frac{1}{2}ft^2$ and $v = u + ft$, find $v$ in terms of $s, f$ and $t$ only.

**5 (Tr).** The formula $S = Vt$ may be used to determine the distance in miles, $S$, covered in $t$ hours by a vehicle travelling at a steady speed of $V$ m.p.h.

Transpose the formula and estimate the time taken, in minutes, by a car travelling at 33 m.p.h. to cover a distance of $9\frac{3}{4}$ miles.

**6 (A).** If $W$ is the live weight of an animal and $w$ the dead weight of saleable meat obtained from it, then

$$w = \frac{4W}{7} \quad \text{for cattle}$$

and

$$w = \frac{4W}{5} \quad \text{for pigs.}$$

(*a*) Estimate the live weight of a bullock that yielded 40 stones of meat.

(*b*) What should be the dead weight of a bacon pig that scaled 1 cwt. 84 lb. when alive?

**7 (T).** The correct diameter, $d$, of rivets to be used for joining boiler plates of thickness $t$ in., is given by the formula

$$d = 1 \cdot 2 \sqrt{t}.$$

Find (*a*) the diameter of rivets suitable for use with plates $\frac{3}{8}$ in. thick (answer to nearest $\frac{1}{100}$th in.); (*b*) what thickness of plates could be joined by rivets of 1 in. diameter? (answer to nearest $\frac{1}{10}$th in.).

[ 124 ]

## FORMULAE

**8 (B).** The number of cubic feet of serviceable timber contained in a tree trunk may be calculated from the formula

$$V = \frac{G^2 L}{2304},$$

where $G$ is the mean girth of the tree in inches (i.e. the average measurement round the trunk taken at regular intervals), and $L$ is the length of the trunk in feet.

(*a*) Find the number of cubic feet of timber in a felled trunk 24 ft. long with girth measurements 72 in., 67 in., 63 in., 62 in., 61 in. at intervals of 6 ft. (answer to nearest cubic foot).

(*b*) Transpose the formula with $G$ as the subject and so determine the mean girth in inches of a tree trunk 18 ft. long with a cubical content of 22·8 cu.ft.

**9 (Tr).** The circumference of an ellipse is found from the formula

$$C = \frac{\pi}{2}(d_1 + d_2),$$

where $d_1$ and $d_2$ are the lengths of the major and minor axes (fig. 95).

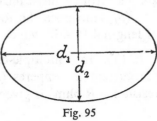

Fig. 95

(*a*) A metal strengthening strap has to be made for a petrol tank, of elliptical cross-section, measuring $12\frac{3}{4}$ in. by $9\frac{1}{2}$ in. What is the length of metal required, allowing 3 in. extra for fastening? (answer to the nearest inch).

(*b*) By transposing the formula find the length of the minor axis of an ellipse with a circumference 2 ft. 6 in. and a major axis $10\frac{1}{2}$ in.

**10** (T). The power in watts ($W$) necessary to bring an electric filament lamp to incandescence (to 'light' the lamp) is obtained from the formula

$$W = I^2R,$$

where $I$ is the current in amperes passing through the filament and $R$ is the resistance of the filament when at incandescent heat.

(*a*) Find $W$ when the filament current is 0·352 amp. and the hot resistance is 323 ohms.

(*b*) Transpose the formula to find the current flowing in the filament of a 60-watt lamp with a hot resistance 666·6 ohms.

**11** (A). The following formula

$$W = \frac{5G^2L}{21},$$

giving $W$ in stones' weight, is used by butchers to estimate the weight of saleable meat, without offal, obtainable from the carcass of a bullock. $G$ is the girth in feet, measured behind the fore-legs and $L$ is the length in feet from shoulder to rump.

(*a*) Estimate the number of stones of meat (to the nearest ½ stone) from a beast measuring 4 ft. 9 in. with a girth of 6 ft.

(*b*) Transpose the formula to find the girth of a bullock of length 4 ft. 6 in. which gave 35½ stones of butchers' meat.

**12** (T). In a multiple-cell battery, when connected in series, the current in amperes ($I$) which flows through a circuit of resistance $R$ ohms is given by the formula

$$I = \frac{nV}{R + nr},$$

where $V$ is the voltage of each cell, $n$ the number of cells in the battery and $r$ the internal resistance of each cell.

# FORMULAE

(a) Estimate the current in a circuit of 3 ohms resistance obtainable from a 10-cell battery, each cell of which has a voltage of 1·5 volts and an internal resistance of 0·2 ohm.

(b) Transpose the formula in order to determine the number of cells there are in a battery which gives a current of 5·45 amp. in a circuit of 2·48 ohms resistance when each cell has a voltage of 1·4 volts and an internal resistance of 0·1 ohm.

**13** (Tr). The R.A.C. rating for the horse-power of a motor engine is obtained from the formula

$$H = \tfrac{2}{5}nd^2,$$

where $d$ is the diameter of each cylinder in inches and $n$ is the number of cylinders.

(a) What is the R.A.C. rating of a four-cylinder car whose cylinders are 3·5 in. in diameter?

(b) What is the diameter of the cylinders of a 25 H.P. six-cylinder marine engine?

**14** (B). The loss of heat in cold weather from a room with one or more 'outside' walls is a consideration in the building of a house. The amount of heat lost, calculated in British Thermal Units ($H$) is found from the formula

$$H = AK(t_1 - t_0),$$

where $K$ is a quantity depending upon the type and thickness of the outside walls, $t_1$ is the indoor temperature, $t_0$ is the outdoor temperature, both in ° F., and $A$ is the area of the outer wall or walls in square feet.

(a) If the outer wall of a room measures 12 ft. by 10 ft. and $K = 0·3$, find the loss of heat when the room temperature is 65° F. and the temperature outdoors is 40° F.

(b) Transpose the formula to find the temperature outdoors if there is a loss of 2490 B.TH.U. from a room at 71° F. with a 9 in. outside brick wall ($K=0\cdot38$) measuring 168 sq.ft.

**15** (Tr). The height in inches ($H$) that the outer rail is raised on a railway curve may be found from the formula

$$H = \frac{4gV^2}{15R},$$

where $g$ is the gauge of the track in feet, $R$ is the radius of the curve in yards and $V$ is the maximum speed in m.p.h. to be used on the curve.

(a) What is the correct height to raise the outer rail of track 4 ft. 8 in. gauge, on a curve of 1500 yd. radius to be safe for speeds up to 60 m.p.h.?

(b) What would be the safe maximum speed of a train on a curve of a 4 ft. 8 in. track whose radius is 860 yd. with the outer rail raised $4\frac{1}{2}$ in.?

**16** (T). When making a nut to fit the Whitworth thread of a bolt, the size of the tap drill for drilling the thread of the nut is found from a formula

$$d = D - \frac{1\cdot28}{n},$$

where $D$ is the diameter of the bolt in inches and $n$ is the number of threads per inch.

(a) If a $\frac{3}{4}$ in. bolt is threaded with 10 threads per in., what should be the diameter of the drill for tapping?

(b) How many threads per inch should there be on a $\frac{7}{8}$ in. bolt when the diameter of the tap drill is 0·733 in.?

[ 128 ]

# CHAPTER 13

# TRIGONOMETRY

So far, in the chapter on areas of plane surfaces, we have considered only the methods employed to determine the area of a triangle from knowledge of the length of the base and the perpendicular drawn to it from the opposite angle. In the study of trigonometry we make use of the angles of a triangle as well as the lengths of the sides. As we are concerned with only the elementary principles we confine all our problems to the right-angled triangle.

## Properties of the Right-angled Triangle

In any right-angled triangle there are certain relationships which are always true and for simplicity of description there are certain names which are always used to denote the various parts of the triangle.

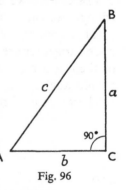

Fig. 96

Let *ABC* (fig. 96) be any right-angled triangle with the right angle at *C*.*

The sides opposite to the various angles of a triangle are given special names. The side opposite the right angle *C* (called *c*) is known as the *hypotenuse* of the triangle. The side opposite the angle *B* (called *b*), i.e. the side upon which the triangle is standing, is known as the *base* of the

* The nomenclature of capital letters for angles and the same small letter to denote the side opposite the corresponding angle is adopted as being the simplest to follow and explain.

[ 129 ]

triangle. And the remaining side $a$ (opposite the angle $A$) is known as the *perpendicular* of the triangle.

We are now able to state a number of facts about a right-angled triangle which are always true no matter what the size of the triangle may be.

(1) In any right-angled triangle the following relationship is always true:
$$a^2 + b^2 = c^2,$$

in other words, if we construct a square with sides equal in length to $a$ and add to it a square with sides equal in length to $b$, then the total area so formed is equal exactly to a square with sides equal in length to $c$. So that

$$c = \sqrt{a^2 + b^2},$$
$$b = \sqrt{c^2 - a^2},$$
and
$$a = \sqrt{c^2 - b^2}.$$

Therefore if we know the length of two sides of a right-angled triangle we can easily find the length of the third side.

(2) Secondly, we know that all the angles of a triangle when added together are equal to *two right angles*. So that when one of them is a right angle the sum of the other two must also be a right angle.

Therefore if angle $C = 1$ right angle (90°). Then

angle $A$ + angle $B = 90°$

and
angle $A = (90° - \text{angle } B),$

also
angle $B = (90° - \text{angle } A).$

(3) Thirdly, in any right-angled triangle, no matter what its size, provided that the corresponding angles are all equal (as in

[ 130 ]

fig. 97 where $\angle A = \angle A'$; $\angle B = \angle B'$; $\angle C = \angle C'$); the following ratios are always true:

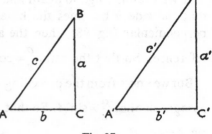

$$\frac{a}{c} = \frac{a'}{c'}; \quad \frac{b}{c} = \frac{b'}{c'}; \quad \frac{a}{b} = \frac{a'}{b'}.$$

In a right-angled triangle the angle opposite the perpendicular, in this case the $\angle A$ (since $a$ is the perpendicular), is called the *angle of reference*, and the above

Fig. 97

ratios have special names applicable to the angle of reference.

The ratio $\frac{a}{c}$ $\left(\text{i.e. } \dfrac{\text{perpendicular}}{\text{hypotenuse}}\right)$ is called the *sine* of the angle of reference, i.e. the ratio $\frac{a}{c} =$ sine $A$.

Similarly the ratio $\frac{b}{c}$ is called *cosine A* and the ratio $\frac{a}{b}$ is called *tangent A*. These ratios, when written, are always abbreviated, so that we write the first three *trigonometrical ratios*, as they are called:

$$\frac{\text{perpendicular}}{\text{hypotenuse}} \left(\frac{a}{c}\right) = \sin A,$$

$$\frac{\text{base}}{\text{hypotenuse}} \left(\frac{b}{c}\right) = \cos A,$$

$$\frac{\text{perpendicular}}{\text{base}} \left(\frac{a}{b}\right) = \tan A.$$

These ratios are the basis of all trigonometrical calculation and should be learned thoroughly before proceeding to the next section.

[ 131 ]

## Complementary Angles

If we examine fig. 96 again and this time we turn it round so that the side $a$ becomes the base and the side $b$ becomes the perpendicular (fig. 98), then the angle $B$ becomes the angle of reference. So that the ratio $\dfrac{a}{c} = \cos B$.

But we know from the preceding paragraph that $\dfrac{a}{c} = \sin A$. So that $\sin A = \cos B$.

Also we know from para. 2 that
$$B = 90° - A.$$

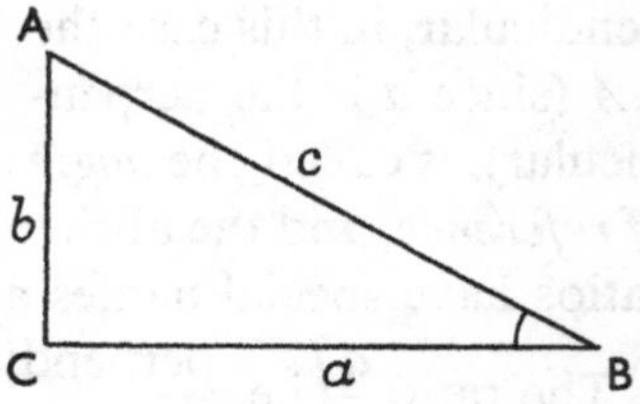

Therefore
$$\sin A = \cos (90° - A)$$

Fig. 98

and, similarly, $\qquad \cos A = \sin (90° - A).$

The angle $(90° - A)$ is called the *complement* of the angle $A$.

Consequently, if we wish to know the value of $\cos A$ we know that it must be the same as $\sin (90° - A)$.

For example, $\qquad \cos 60° = \sin (90° - 60°)$
$$= \sin 30°$$
$$= 0·5.$$

(Show that $\sin 45° = \cos 45°$.)

Therefore in the tables at the end of the book for finding the value of the sine of any angle between 0° and 90° we can use the same table *backwards* to read the values of cosines, the only difference being that the sizes of the angles are written from top to bottom (on the left) for the sines and from bottom to top (on the right) for the cosines.

[ 132 ]

Consider fig. 98 again, where we know that

$$\frac{b}{a} = \tan B = \tan (90° - A).$$

We also know that $\quad \dfrac{b}{a} = \dfrac{1}{\dfrac{a}{b}} = \dfrac{1}{\tan A}.$

Therefore $\quad\quad \tan (90° - A) = \dfrac{1}{\tan A}$

and, similarly, $\quad \tan A = \dfrac{1}{\tan (90° - A)}.$

We are frequently called upon to use the ratio $\dfrac{1}{\tan A}$ so it is given a special name—*cotangent A* (written cot *A*). Consequently the tangent tables, also at the end of the book, may be read backwards (as in the case of the sine and cosine tables) for finding values of the cotangent. (The reciprocals $\dfrac{1}{\sin A}$ and $\dfrac{1}{\cos A}$, although not so important, also have special names which are *cosecant A* (cosec *A*) and *secant A* (sec *A*) respectively.)

Trigonometry has many uses and enables us easily to work such problems as the measurement of inaccessible heights or distances (it is not necessary to climb a mountain to find its height). The science of navigation at sea or in the air owes much to trigonometry and its application for its accuracy. By using the trigonometrical ratios of a right-angled triangle and by careful measurement of angles by means of special instruments (theodolite or sextant) we are able to determine heights and distances quite accurately. Much interest therefore may be derived from the first eight questions of the following exercise (No. 16) in practical trigonometry before proceeding with the succeeding problems. For this purpose the following primitive measuring instruments will be useful and most of them can easily be made:

(1) A 30 ft. or 60 ft. surveyors' tape measure.

(2) A large plywood or cardboard protractor. This should measure about 1 ft. in radius and should be numbered in degrees, from a smaller protractor, as shown in fig. 99.

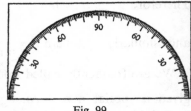

(3) Two or three long thin battens, about 6 ft. in length.

(4) A simple instrument (called a clinometer) for measuring angles of height (*inclination*) and angles of depth (*depression*) when objects are above or below the level of the eye. This may be easily constructed as in fig. 100.

Fig. 99

(5) A right-angled square with sides about 3 ft. long (fig. 101).

If necessary this last item may be dispensed with, although it is useful for quickness.

A right angle on the ground can always be constructed in this way.

We know that in any right-angled triangle (with the right angle at $C$)

$$a^2 + b^2 = c^2.$$

So that if we make $a = 3$ ft. and $b = 4$ ft. we have

$$c^2 = 3^2 + 4^2$$

$$= 9 + 16 = 25 \text{ sq.ft.}$$

So that $c = 5$ ft.

[ 134 ]

Consequently, to construct a line at right angles to any other straight line from a point $A$ (fig. 102) we first measure from $A$

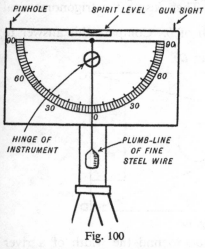

Fig. 100

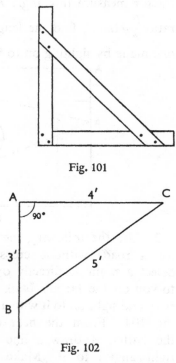

Fig. 101

Fig. 102

along the line a distance $AB = 3$ ft. Fix the tape measure with the end (zero) at $A$ and the 9 ft. mark at $B$. Hold the tape at the 4 ft. mark and pull it so that both lengths are taut (measuring 4 ft. and 5 ft. respectively). The point $C$, so obtained, when joined to $A$ will give us a line at right angles to $AB$.

## EXERCISE 16

**1.** The length of a full-sized football pitch is 120 yd. Check this measurement by measuring from a corner flag $A$ a distance of 60 ft. (20 yd.) along the goal line. At this point $B$ 'sight' on to the corresponding corner flag $C$ at the other end of the pitch by lining up one of the battens (fig. 103). With the large protractor measure this angle $B$. Then, using the trigonometrical ratio $\dfrac{b}{c} = \tan B$, find the length of $b$. Compare this answer to one made by sighting on to flag $D$.

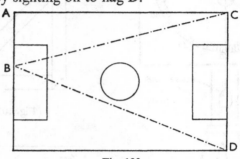

Fig. 103

2. Use the following method to find the width of a river (or a road) without crossing it. Select a point $X$ directly opposite to you on the farther bank of the river and sight on to it with a batten (fig. 104). From the near end of the batten $A$ draw a line $AB$ at right angles to it. Make $AB$ a known length, say 30 ft. At $B$ sight on to the same point $X$ with another batten and measure the angle $B$ with the protractor. Again using the

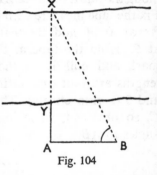

Fig. 104

tangent ratio for the angle $B$, find the length of $XA$. Measure the length $AY$ and subtract this from $XA$ to give you the width of the river.

[ 136 ]

**3.** Kick a football from a fixed point on the playing field and find the length of your kick without direct measurement.

**4.** Measure the height of a tree or a telephone post by using the clinometer as shown in fig. 105. Remember to *add* the height of the instrument (4 ft.) to your answer.

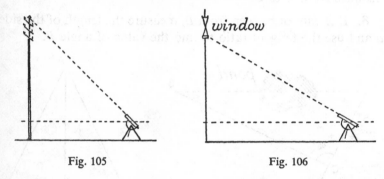

Fig. 105          Fig. 106

**5.** Measure the height of a window-sill or balcony in the upper part of the school building by angular measurement from a point on the ground at a known distance from the wall vertically below it (fig. 106).

**6.** Check this height by measuring the angle of depression from the window or balcony on to the same spot. (Remember to allow for the height of instrument in each case, either adding or subtracting as necessary.)

**7.** If possible select two objects, or arrange two stakes, in such a position that there is an obstruction in between them. Such as two trees with a pond between them (fig. 107). Use the following method to measure how far they are apart.

Fix the end of the tape measure at the point *A* and stretch the tape along any convenient line *AX*. With one edge of the

[ 137 ]

square upon the line *AX* slide this along to the point *C* where the other edge is sighted on to the point *B*.

Note the length of *AC* along the tape. With the protractor measure the angle *A*. Then from the ratio $\dfrac{b}{c} = \cos A$ find the distance from *A* to *B*.

**8.** If *A* cannot be seen from *B*, measure the length of the side *a* and use the tangent ratio to find the value of angle *B*.

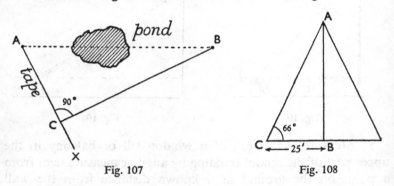

Fig. 107          Fig. 108

**9** (G). A mast *AB* is stayed to the ground by wire stays, as shown in fig. 108. If the distance from the end of one stay *C* to the base of the mast is 25 ft. and the angle *C* is found to be 66°, what is the height of the mast?

**10** (B). When a road runs uphill or downhill there is a continual difference in height of the road surface as we travel along it. The ratio of distance travelled along the road for an alteration of 1 ft. in height is called the *gradient* of the slope. Thus, if for every 1 ft. rise in height we travel 8 ft. along the road surface, then the gradient is 1 in 8, or $\frac{1}{8}$.

(*a*) What is the angle of slope of a steep road rising 1 in 4?

[ 138 ]

(*b*) What length of road surface must be allowed for when re-metalling the surface of a hill with a gradient 1 in 20, if the height of the foot of the hill is 250 ft. and the height of the top of the hill is 425 ft. above sea-level?

**11 (T).** What is the depth, to the nearest thousandth of an inch, of a V thread when cut with five threads to the inch?

*Note.* In a V thread the pitch and the sides of the thread form the sides of an equilateral triangle.

**12 (Tr).** On a map of scale 1 in. = 1 mile a length of straight road crosses the 150 ft. contour line at *A* and the 350 ft. contour line at *B*. The distance from *A* to *B* on the map is $2\frac{1}{2}$ in. Find (*a*) the angle of slope of the road, (*b*) the gradient.

**13 (A).** The triangle *ABC* (fig. 109) represents a three-sided field. Its area may be determined from the measurements of any two sides and the angle included between them. For example, if the two sides *b* and *c* and the included angle *A* are known, then the area of the field is $\frac{1}{2}bc \sin A$. Use logarithms to find what would be the cost of manuring such a field at the rate of £3 per acre when the measurements are: *b* = 150 yd., *c* = 208 yd., *A* = 55°.

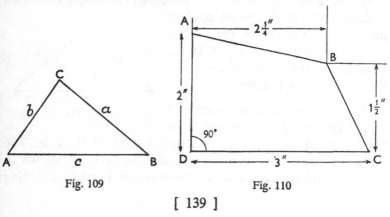

Fig. 109    Fig. 110

**14** (T).  A flat metal plate is dimensioned as shown in fig. 110. Find the value of angles $A$ and $C$.

Without working further trigonometrical ratios, find the value of angle $B$.

**15** (B).  To determine the height of a chimney stack a theodolite is set up at a distance of 135 ft. from the base of the stack. The height of the instrument is 4 ft. 6 in. above ground-level. If the angle of inclination to the top of the stack is found to be 43° 15′, what is the height of the stack?

**16** (Tr).  A seaplane mooring buoy is to be anchored at a distance of 250 yd. to seaward of a vertical cliff 171 ft. high. What will be the angle of depression to the buoy from the nearest point on the cliff top when the buoy is in correct position?

**17** (G).  A navigator in an aircraft flying at a steady altitude of 10,000 ft. reads the angle of depression to an aerodrome in the direct line of flight as 23° 30′.  1 min. 10 sec. later the aircraft is vertically above the aerodrome. What is the speed of the aircraft in miles per hour?  (Use logarithms.)

**18** (B).  The diagram in fig. 111 represents a simple roof-truss in timber. The angle $P$ is the pitch of the roof. If the rise of the roof is 10 ft. 6 in. and the span is 45 ft., what is (*a*) the pitch of the roof, (*b*) the length of a rafter?

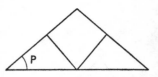

Fig. 111

**19** (T).  An overhead electrical supply line has to be carried from point $A$ to $B$ and back to $C$ across a road 32 ft. wide. The angles so formed between the line and the kerb are shown in fig. 112.  Find (*a*) the length of cable between $A$ and $C$, assuming the height is constant; (b) the distance from A to C. (Answer to nearest foot.)

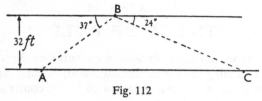

Fig. 112

**20 (A).** A line of pipes for field drainage are to be laid in a field of uniform gradient between points A and B (fig. 113). In order to determine the 'fall' (or gradient) of the proposed drainage the following simple method was adopted.

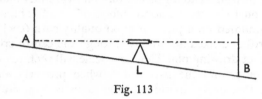

Fig. 113

A surveyor's staff (or a long rod marked in feet and decimals of a foot) was set up at A, and at a convenient distance between A and B a 'dumpy' level was placed. This instrument is simply a telescope on a tripod so arranged that it can be accurately 'levelled' and can rotate through any angle horizontally. The level was focused on to the staff at A and the reading taken. The staff was then set up at B and a similar reading taken on the staff.

Suppose that these readings were 1·1 ft. and 13·8 ft. respectively. The distance from A to B was measured as 124 yd. (a) What will be the fall of the drain if the head at A is 2 ft. below surface and the end at B is 3 ft. 6 in. below surface? (b) What will be the angle of slope?

[ 141 ]

# CHAPTER 14

# THE SLIDE RULE

Logarithms, as we have already seen, are a valuable short cut to any calculation involving multiplication, division, square, square root, etc., and the same principle has been used in the construction of an instrument called the slide rule.

Many slide rules are complicated and specialized in their design but in principle they owe their usefulness to the same method of construction.

Printed on the end-papers of this book are two circular scales, the smaller one of which should be cut out very carefully and the larger one pasted on to a small smooth block of wood. The smaller one should be mounted on a piece of good quality card (such as 3-sheet Bristol board) and fixed accurately through the centre cross of each scale by a stout drawing pin. The inner scale will then revolve exactly within the outer and the two scales, when properly aligned, will coincide, so that every division on the inner is opposite the corresponding division on the outer.

The scales themselves have been so constructed that the distance around the circumference between any two numbers (e.g. 3 and 2) is proportional to the difference between their *logarithms*, i.e. the distance is proportional to

$$\log 3 - \log 2 = 0.4771 - 0.3010$$
$$= 0.1761.$$

Similarly the distance from the unity mark 1 to 2 is proportional to

$$\log 2 - \log 1 = 0.3010 - 0$$
$$= 0.3010$$

and so on.

Now, by experimenting with a few simple exercises in multiplication and division, let us see how the slide rule is employed.

## Multiplication

*Example* 1. What is the value of $12 \times 7$?

Turn the unity mark on the inner scale until it coincides with 12 on the outer.

Now, opposite 7 on the inner scale we read 84 on the outer.

Satisfy yourself that the same result is obtained by turning the unity mark on the inner scale to 7 on the outer and then reading the value opposite 12 on the inner.

In most cases, with a primitive instrument such as this, decimal values have to be estimated 'by eye'. For this reason accuracy will be only approximate but there should seldom be an error of more than $0.5 \%$.

Practise, first of all, with simple numbers such as $3 \times 5$, $6 \times 8$, $4 \times 11$, etc., before attempting anything harder or numbers involving decimals.

So far we have confined ourselves to numbers below 100, although the slide rule may be used for 100's or 1000's when required, in precisely the same way.

We know that

$$\log 300 - \log 200 = 2 \cdot 4771 - 2 \cdot 3010$$
$$= 0 \cdot 1761$$

which is the same answer as we obtained from $\log 3 - \log 2$. So that the divisions marked 3 and 2 on the scale may equally well be called 300 and 200 (or 3000 and 2000), provided that the rest of the scale is taken as conforming to the same multiple.

## Division

*Example* 2. What is the value of $153 \div 17$?

This time turn the inner scale until 17 is opposite 153 on the outer. (Be careful in the selection of this point.)

Then, on the outer scale, opposite the unity mark, read 9.

## Combined Multiplication and Division

*Example* 3. Find the value of $\dfrac{12\cdot6\times3\cdot4}{4\cdot8}$.

With the slide rule such problems as this can be done in one simple operation.

First divide 12·6 by 4·8 by turning the inner scale so that 4·8 is opposite 12·6 on the outer. The unity mark on the inner scale is then opposite the number which we have to multiply by 3·4.

By leaving the scale in this position and taking the outer reading opposite 3·4 on the inner we have the required answer (by estimation) 8·95.

Make quite certain that this process is properly understood because this is one simple illustration of the many ways in which the slide rule can be a quick and reliable time saver.

In order to gain practice and confidence in using the slide rule the following exercise is given. As many more examples of the same sort may be worked as necessary.

(*Note*. The answers given in the answer book are worked on the slide rule and may not be exactly in agreement if worked arithmetically.)

### EXERCISE 17

**1.** Find the value of (*a*) $16\times2\cdot8$, (*b*) $14\cdot5\times3\cdot9$, (*c*) $8\cdot6\times10\cdot2$, (*d*) $1\cdot25\times12\cdot7$, (*e*) $25\cdot3\times4\cdot8$.

**2.** Find the decimal value for $\pi$ assuming that it is equal to $22\div7$.

**3.** Find the value of (*a*) $23\cdot5\div16$, (*b*) $110\div14\cdot6$, (*c*) $258\div17\cdot2$, (*d*) $149\div86\cdot6$, (*e*) $1\cdot75\div2\cdot4$.

**4.** By using one operation with the slide rule find the value of

(*a*) $\dfrac{2\cdot9\times13\cdot4}{3\cdot3}$, (*b*) $\dfrac{26\cdot25\times18\cdot1}{32\cdot4}$, (*c*) $\dfrac{96\times1\cdot55}{13\cdot8}$, (*d*) $\dfrac{152\cdot5\times16\cdot7}{94\cdot5}$,

(*e*) $\dfrac{258\times5\cdot65}{24\cdot7}$.

[ 144 ]

**5.** By using only two operations with the slide rule find the value of

(a) $\dfrac{3\cdot8\times41\cdot6}{12\cdot2\times4\cdot5}$, (b) $\dfrac{62\cdot5\times88\cdot8}{11\cdot7\times25}$, (c) $\dfrac{252\times180}{23\cdot6\times59\cdot5}$, (d) $\dfrac{(14\cdot8)^2}{8\times5\cdot1}$, (e) $\left(\dfrac{61\cdot5}{18\cdot3}\right)^2$.

## Conversions

Another useful calculation that can easily be made on the slide rule is the conversion of a quantity expressed in terms of one unit into its corresponding value in another.

*Example* 4. What is the capacity in cubic feet of a tank holding 250 gal.?

We know that 1 cu.ft. is equivalent to 6·25 gal. So that if the unity mark on the inner scale is turned to 6·25 on the outer, then every reading on the inner scale, in cubic feet, is opposite the corresponding gallonage on the outer scale and vice versa.

The reading opposite 250 gal. on the outer scale is 40 cu.ft. on the inner.

In this way as many different conversions as required can be made provided we have one definite known ratio between the two different units.

The scale is marked already for a few conversions.

## Speed Conversion

In the same way, if we wish to know the distance covered in any interval of time by a body moving at a known speed it can readily be found from the slide rule.

*Example* 5. How far will an aircraft travel in 18 min. at a steady ground speed of 355 m.p.h.?

If we turn the arrow (hr.) on the 60 mark of the inner scale (representing 60 min.) to the distance flown in 1 hr. (355 miles), then all divisions on the inner scale treated as minutes are opposite the corresponding distance flown, and the answer in this case is 107 miles.

*Example* 6. What is the speed in miles per hour of a car which covers a distance of 32·5 miles in 26·3 min.?

By turning the inner scale so that 26·3 (min.) is opposite 32·5 (miles) on the outer, we read opposite the hr. mark—74 m.p.h.

The above examples are a few of the types of simple problem that can be 'answered' by the slide rule.

The following exercise, in which a little ingenuity is sometimes called for, affords good practice in the use of the instrument and illustrates all the foregoing principles.

## EXERCISE 18

**1.** Find the gallonage of the following volumes which are in cubic feet, (*a*) 20, (*b*) 17, (*c*) 4·55, (*d*) 1·27, (*e*) 63.

**2.** What are the volumes in cubic feet of tanks having the following capacities in gallons: (*a*) 100, (*b*) 250, (*c*) 12·5, (*d*) 8, (*e*) 2000?

**3.** Convert the following lengths in feet into their equivalents in metres: (*a*) 29, (*b*) 100, (*c*) 325·5, (*d*) 8·25, (*e*) 2·5.

**4.** If the height of Mt. Everest is given as 8884·4 m., what is its approximate height in feet?

**5.** Find the area in square feet of panes of glass measuring (*a*) 4 ft. 3 in. by 2 ft. 6 in., (*b*) 2 ft. 9 in. by 3 ft. 3 in., (*c*) 3 ft. 6 in. square, (*d*) 1 ft. 4 in. by 5 ft. 8 in., (*e*) 9 in. by 40 in.

**6.** What distance, in each case, will be covered in the specified time by a car travelling as follows: (*a*) 56 m.p.h. for 8 min., (*b*) 64 m.p.h. for 7 min., (*c*) 72 m.p.h. for 11·3 min., (*d*) 82·5 m.p.h. for 3 min. 30 sec., (*e*) 48 m.p.h. for 10 min. 18 sec.?

**7.** Convert into miles per hour the following recorded test flights: (*a*) 13 miles in 3·3 min., (*b*) 28·5 miles in 4·2 min., (*c*) 87 miles in 15·75 min., (*d*) 46·5 miles in 12 min. 15 sec., (*e*) 11·6 miles in 2 min. 48 sec.

**8.** The petrol gauge on a motor car reads 6¾ gal. as being the amount of petrol in the tank. What is the approximate mileage that could be covered without refuelling if the average petrol consumption of the car is 29·5 miles per gallon?

**9.** A lorry on test covered 182·3 miles on 18·6 gal. of petrol. What does this represent as a fuel consumption in miles per gallon?

**10.** What is the speed in miles per hour of an aircraft which covered a measured distance of 2 miles in 24·5 sec.?

*Note.* Try to work all the remaining problems with only *one* operation of the slide rule.

**11.** A certain artificial manure is made up so that 1 cwt. contains 42 lb. of nitrate of soda. What is the percentage of nitrate of soda in the manure?

**12.** An artificial fertilizer is being made up from a formula which includes 3·5 % of potash. What weight of potash, in pounds, should there be in 3 cwt. of the fertilizer?

**13.** The electrical resistance of a certain wire is known to be 2·16 ohms per 100 yd. What is the resistance of 320 yd. of this wire?

**14.** The resistance of another wire is given as 27 ohms per mile. What is the resistance of this wire per 100 yd.?

**15.** If the resistance of a section of conductor rail measuring 271 yd. is found to be 0·35 ohm, what is the probable resistance per mile?

**16.** A map is drawn to a scale of 2½ in. to the mile. How far is it from Preston to Blackpool if the distance on this map measures 3 ft. 6¾ in.?

**17.** Two landmarks shown on a map measure 1·94 in. apart. They are known to be 29 miles apart. What is the scale of the map in miles to the inch?

**18.** What is the cross-sectional area, in each case, of a tube with internal bore measuring: (*a*) 2 in., (*b*) 3 in., (*c*) 6 in., (*d*) 1·2 in., (*e*) 0·5 in.?

**19.** What is the approximate cost of preparing and concreting a yard measuring 140 ft. by 86 ft. if the labour and material are estimated to work out at 6*d.* per square foot?

**20.** A field is to be enclosed by a fence made of split timber palings laid at the rate of 7 palings per yard length. If the area to be enclosed is a rectangle measuring 120 yd. by 55 yd., what will be the cost of the palings if they are bought in bundles, each bundle averaging 38 palings and costing £1?

# LOGARITHMS OF NUMBERS

| | 0 | 1 | 2 | 3 | 4 | 5 | 6 | 7 | 8 | 9 | 1 | 2 | 3 | 4 | 5 | 6 | 7 | 8 | 9 |
|---|---|---|---|---|---|---|---|---|---|---|---|---|---|---|---|---|---|---|---|
| 10 | 0000 | 0043 | 0086 | 0128 | 0170 | 0212 | 0253 | 0294 | 0334 | 0374 | 4 | 8 | 12 | 17 | 21 | 25 | 29 | 33 | 37 |
| 11 | 0414 | 0453 | 0492 | 0531 | 0569 | 0607 | 0645 | 0682 | 0719 | 0755 | 4 | 8 | 11 | 15 | 19 | 23 | 26 | 30 | 34 |
| 12 | 0792 | 0828 | 0864 | 0899 | 0934 | 0969 | 1004 | 1038 | 1072 | 1106 | 3 | 7 | 10 | 14 | 17 | 21 | 24 | 28 | 31 |
| 13 | 1139 | 1173 | 1206 | 1239 | 1271 | 1303 | 1335 | 1367 | 1399 | 1430 | 3 | 6 | 10 | 13 | 16 | 19 | 23 | 26 | 29 |
| 14 | 1461 | 1492 | 1523 | 1553 | 1584 | 1614 | 1644 | 1673 | 1703 | 1732 | 3 | 6 | 9 | 12 | 15 | 18 | 21 | 24 | 27 |
| 15 | 1761 | 1790 | 1818 | 1847 | 1875 | 1903 | 1931 | 1959 | 1987 | 2014 | 3 | 6 | 8 | 11 | 14 | 17 | 20 | 22 | 25 |
| 16 | 2041 | 2068 | 2095 | 2122 | 2148 | 2175 | 2201 | 2227 | 2253 | 2279 | 3 | 5 | 8 | 11 | 13 | 16 | 18 | 21 | 24 |
| 17 | 2304 | 2330 | 2355 | 2380 | 2405 | 2430 | 2455 | 2480 | 2504 | 2529 | 2 | 5 | 7 | 10 | 12 | 15 | 17 | 20 | 22 |
| 18 | 2553 | 2577 | 2601 | 2625 | 2648 | 2672 | 2695 | 2718 | 2742 | 2765 | 2 | 5 | 7 | 9 | 12 | 14 | 16 | 19 | 21 |
| 19 | 2788 | 2810 | 2833 | 2856 | 2878 | 2900 | 2923 | 2945 | 2967 | 2989 | 2 | 4 | 7 | 9 | 11 | 13 | 16 | 18 | 20 |
| 20 | 3010 | 3032 | 3054 | 3075 | 3096 | 3118 | 3139 | 3160 | 3181 | 3201 | 2 | 4 | 6 | 8 | 11 | 13 | 15 | 17 | 19 |
| 21 | 3222 | 3243 | 3263 | 3284 | 3304 | 3324 | 3345 | 3365 | 3385 | 3404 | 2 | 4 | 6 | 8 | 10 | 12 | 14 | 16 | 18 |
| 22 | 3424 | 3444 | 3464 | 3483 | 3502 | 3522 | 3541 | 3560 | 3579 | 3598 | 2 | 4 | 6 | 8 | 10 | 12 | 14 | 15 | 17 |
| 23 | 3617 | 3636 | 3655 | 3674 | 3692 | 3711 | 3729 | 3747 | 3766 | 3784 | 2 | 4 | 6 | 7 | 9 | 11 | 13 | 15 | 17 |
| 24 | 3802 | 3820 | 3838 | 3856 | 3874 | 3892 | 3909 | 3927 | 3945 | 3962 | 2 | 4 | 5 | 7 | 9 | 11 | 12 | 14 | 16 |
| 25 | 3979 | 3997 | 4014 | 4031 | 4048 | 4065 | 4082 | 4099 | 4116 | 4133 | 2 | 3 | 5 | 7 | 9 | 10 | 12 | 14 | 15 |
| 26 | 4150 | 4166 | 4183 | 4200 | 4216 | 4232 | 4249 | 4265 | 4281 | 4298 | 2 | 3 | 5 | 7 | 8 | 10 | 11 | 13 | 15 |
| 27 | 4314 | 4330 | 4346 | 4362 | 4378 | 4393 | 4409 | 4425 | 4440 | 4456 | 2 | 3 | 5 | 6 | 8 | 9 | 11 | 13 | 14 |
| 28 | 4472 | 4487 | 4502 | 4518 | 4533 | 4548 | 4564 | 4579 | 4594 | 4609 | 2 | 3 | 5 | 6 | 8 | 9 | 11 | 12 | 14 |
| 29 | 4624 | 4639 | 4654 | 4669 | 4683 | 4698 | 4713 | 4728 | 4742 | 4757 | 1 | 3 | 4 | 6 | 7 | 9 | 10 | 12 | 13 |
| 30 | 4771 | 4786 | 4800 | 4814 | 4829 | 4843 | 4857 | 4871 | 4886 | 4900 | 1 | 3 | 4 | 6 | 7 | 9 | 10 | 11 | 13 |
| 31 | 4914 | 4928 | 4942 | 4955 | 4969 | 4983 | 4997 | 5011 | 5024 | 5038 | 1 | 3 | 4 | 6 | 7 | 8 | 10 | 11 | 12 |
| 32 | 5051 | 5065 | 5079 | 5092 | 5105 | 5119 | 5132 | 5145 | 5159 | 5172 | 1 | 3 | 4 | 5 | 7 | 8 | 9 | 11 | 12 |
| 33 | 5185 | 5198 | 5211 | 5224 | 5237 | 5250 | 5263 | 5276 | 5289 | 5302 | 1 | 3 | 4 | 5 | 6 | 8 | 9 | 10 | 12 |
| 34 | 5315 | 5328 | 5340 | 5353 | 5366 | 5378 | 5391 | 5403 | 5416 | 5428 | 1 | 3 | 4 | 5 | 6 | 8 | 9 | 10 | 11 |
| 35 | 5441 | 5453 | 5465 | 5478 | 5490 | 5502 | 5514 | 5527 | 5539 | 5551 | 1 | 2 | 4 | 5 | 6 | 7 | 9 | 10 | 11 |
| 36 | 5563 | 5575 | 5587 | 5599 | 5611 | 5623 | 5635 | 5647 | 5658 | 5670 | 1 | 2 | 4 | 5 | 6 | 7 | 8 | 10 | 11 |
| 37 | 5682 | 5694 | 5705 | 5717 | 5729 | 5740 | 5752 | 5763 | 5775 | 5786 | 1 | 2 | 3 | 5 | 6 | 7 | 8 | 9 | 10 |
| 38 | 5798 | 5809 | 5821 | 5832 | 5843 | 5855 | 5866 | 5877 | 5888 | 5899 | 1 | 2 | 3 | 5 | 6 | 7 | 8 | 9 | 10 |
| 39 | 5911 | 5922 | 5933 | 5944 | 5955 | 5966 | 5977 | 5988 | 5999 | 6010 | 1 | 2 | 3 | 4 | 5 | 7 | 8 | 9 | 10 |
| 40 | 6021 | 6031 | 6042 | 6053 | 6064 | 6075 | 6085 | 6096 | 6107 | 6117 | 1 | 2 | 3 | 4 | 5 | 6 | 8 | 9 | 10 |
| 41 | 6128 | 6138 | 6149 | 6160 | 6170 | 6180 | 6191 | 6201 | 6212 | 6222 | 1 | 2 | 3 | 4 | 5 | 6 | 7 | 8 | 9 |
| 42 | 6232 | 6243 | 6253 | 6263 | 6274 | 6284 | 6294 | 6304 | 6314 | 6325 | 1 | 2 | 3 | 4 | 5 | 6 | 7 | 8 | 9 |
| 43 | 6335 | 6345 | 6355 | 6365 | 6375 | 6385 | 6395 | 6405 | 6415 | 6425 | 1 | 2 | 3 | 4 | 5 | 6 | 7 | 8 | 9 |
| 44 | 6435 | 6444 | 6454 | 6464 | 6474 | 6484 | 6493 | 6503 | 6513 | 6522 | 1 | 2 | 3 | 4 | 5 | 6 | 7 | 8 | 9 |
| 45 | 6532 | 6542 | 6551 | 6561 | 6571 | 6580 | 6590 | 6599 | 6609 | 6618 | 1 | 2 | 3 | 4 | 5 | 6 | 7 | 8 | 9 |
| 46 | 6628 | 6637 | 6646 | 6656 | 6665 | 6675 | 6684 | 6693 | 6702 | 6712 | 1 | 2 | 3 | 4 | 5 | 6 | 7 | 7 | 8 |
| 47 | 6721 | 6730 | 6739 | 6749 | 6758 | 6767 | 6776 | 6785 | 6794 | 6803 | 1 | 2 | 3 | 4 | 5 | 5 | 6 | 7 | 8 |
| 48 | 6812 | 6821 | 6830 | 6839 | 6848 | 6857 | 6866 | 6875 | 6884 | 6893 | 1 | 2 | 3 | 4 | 4 | 5 | 6 | 7 | 8 |
| 49 | 6902 | 6911 | 6920 | 6928 | 6937 | 6946 | 6955 | 6964 | 6972 | 6981 | 1 | 2 | 3 | 4 | 4 | 5 | 6 | 7 | 8 |
| 50 | 6990 | 6998 | 7007 | 7016 | 7024 | 7033 | 7042 | 7050 | 7059 | 7067 | 1 | 2 | 3 | 3 | 4 | 5 | 6 | 7 | 8 |
| 51 | 7076 | 7084 | 7093 | 7101 | 7110 | 7118 | 7126 | 7135 | 7143 | 7152 | 1 | 2 | 3 | 3 | 4 | 5 | 6 | 7 | 8 |
| 52 | 7160 | 7168 | 7177 | 7185 | 7193 | 7202 | 7210 | 7218 | 7226 | 7235 | 1 | 2 | 2 | 3 | 4 | 5 | 6 | 7 | 7 |
| 53 | 7243 | 7251 | 7259 | 7267 | 7275 | 7284 | 7292 | 7300 | 7308 | 7316 | 1 | 2 | 2 | 3 | 4 | 5 | 6 | 6 | 7 |
| 54 | 7324 | 7332 | 7340 | 7348 | 7356 | 7364 | 7372 | 7380 | 7388 | 7396 | 1 | 2 | 2 | 3 | 4 | 5 | 6 | 6 | 7 |

# LOGARITHMS OF NUMBERS

Differences

| | 0 | 1 | 2 | 3 | 4 | 5 | 6 | 7 | 8 | 9 | 1 | 2 | 3 | 4 | 5 | 6 | 7 | 8 | 9 |
|---|---|---|---|---|---|---|---|---|---|---|---|---|---|---|---|---|---|---|---|
| 55 | 7404 | 7412 | 7419 | 7427 | 7435 | 7443 | 7451 | 7459 | 7466 | 7474 | 1 | 2 | 2 | 3 | 4 | 5 | 5 | 6 | 7 |
| 56 | 7482 | 7490 | 7497 | 7505 | 7513 | 7520 | 7528 | 7536 | 7543 | 7551 | 1 | 2 | 2 | 3 | 4 | 5 | 5 | 6 | 7 |
| 57 | 7559 | 7566 | 7574 | 7582 | 7589 | 7597 | 7604 | 7612 | 7619 | 7627 | 1 | 2 | 2 | 3 | 4 | 5 | 5 | 6 | 7 |
| 58 | 7634 | 7642 | 7649 | 7657 | 7664 | 7672 | 7679 | 7686 | 7694 | 7701 | 1 | 1 | 2 | 3 | 4 | 4 | 5 | 6 | 7 |
| 59 | 7709 | 7716 | 7723 | 7731 | 7738 | 7745 | 7752 | 7760 | 7767 | 7774 | 1 | 1 | 2 | 3 | 4 | 4 | 5 | 6 | 7 |
| 60 | 7782 | 7789 | 7796 | 7803 | 7810 | 7818 | 7825 | 7832 | 7839 | 7846 | 1 | 1 | 2 | 3 | 4 | 4 | 5 | 6 | 6 |
| 61 | 7853 | 7860 | 7868 | 7875 | 7882 | 7889 | 7896 | 7903 | 7910 | 7917 | 1 | 1 | 2 | 3 | 4 | 4 | 5 | 6 | 6 |
| 62 | 7924 | 7931 | 7938 | 7945 | 7952 | 7959 | 7966 | 7973 | 7980 | 7987 | 1 | 1 | 2 | 3 | 3 | 4 | 5 | 6 | 6 |
| 63 | 7993 | 8000 | 8007 | 8014 | 8021 | 8028 | 8035 | 8041 | 8048 | 8055 | 1 | 1 | 2 | 3 | 3 | 4 | 5 | 5 | 6 |
| 64 | 8062 | 8069 | 8075 | 8082 | 8089 | 8096 | 8102 | 8109 | 8116 | 8122 | 1 | 1 | 2 | 3 | 3 | 4 | 5 | 5 | 6 |
| 65 | 8129 | 8136 | 8142 | 8149 | 8156 | 8162 | 8169 | 8176 | 8182 | 8189 | 1 | 1 | 2 | 3 | 3 | 4 | 5 | 5 | 6 |
| 66 | 8195 | 8202 | 8209 | 8215 | 8222 | 8228 | 8235 | 8241 | 8248 | 8254 | 1 | 1 | 2 | 3 | 3 | 4 | 5 | 5 | 6 |
| 67 | 8261 | 8267 | 8274 | 8280 | 8287 | 8293 | 8299 | 8306 | 8312 | 8319 | 1 | 1 | 2 | 3 | 3 | 4 | 5 | 5 | 6 |
| 68 | 8325 | 8331 | 8338 | 8344 | 8351 | 8357 | 8363 | 8370 | 8376 | 8382 | 1 | 1 | 2 | 3 | 3 | 4 | 4 | 5 | 6 |
| 69 | 8388 | 8395 | 8401 | 8407 | 8414 | 8420 | 8426 | 8432 | 8439 | 8445 | 1 | 1 | 2 | 2 | 3 | 4 | 4 | 5 | 6 |
| 70 | 8451 | 8457 | 8463 | 8470 | 8476 | 8482 | 8488 | 8494 | 8500 | 8506 | 1 | 1 | 2 | 2 | 3 | 4 | 4 | 5 | 6 |
| 71 | 8513 | 8519 | 8525 | 8531 | 8537 | 8543 | 8549 | 8555 | 8561 | 8567 | 1 | 1 | 2 | 2 | 3 | 4 | 4 | 5 | 5 |
| 72 | 8573 | 8579 | 8585 | 8591 | 8597 | 8603 | 8609 | 8615 | 8621 | 8627 | 1 | 1 | 2 | 2 | 3 | 4 | 4 | 5 | 5 |
| 73 | 8633 | 8639 | 8645 | 8651 | 8657 | 8663 | 8669 | 8675 | 8681 | 8686 | 1 | 1 | 2 | 2 | 3 | 4 | 4 | 5 | 5 |
| 74 | 8692 | 8698 | 8704 | 8710 | 8716 | 8722 | 8727 | 8733 | 8739 | 8745 | 1 | 1 | 2 | 2 | 3 | 4 | 4 | 5 | 5 |
| 75 | 8751 | 8756 | 8762 | 8768 | 8774 | 8779 | 8785 | 8791 | 8797 | 8802 | 1 | 1 | 2 | 2 | 3 | 3 | 4 | 5 | 5 |
| 76 | 8808 | 8814 | 8820 | 8825 | 8831 | 8837 | 8842 | 8848 | 8854 | 8859 | 1 | 1 | 2 | 2 | 3 | 3 | 4 | 5 | 5 |
| 77 | 8865 | 8871 | 8876 | 8882 | 8887 | 8893 | 8899 | 8904 | 8910 | 8915 | 1 | 1 | 2 | 2 | 3 | 3 | 4 | 4 | 5 |
| 78 | 8921 | 8927 | 8932 | 8938 | 8943 | 8949 | 8954 | 8960 | 8965 | 8971 | 1 | 1 | 2 | 2 | 3 | 3 | 4 | 4 | 5 |
| 79 | 8976 | 8982 | 8987 | 8993 | 8998 | 9004 | 9009 | 9015 | 9020 | 9025 | 1 | 1 | 2 | 2 | 3 | 3 | 4 | 4 | 5 |
| 80 | 9031 | 9036 | 9042 | 9047 | 9053 | 9058 | 9063 | 9069 | 9074 | 9079 | 1 | 1 | 2 | 2 | 3 | 3 | 4 | 4 | 5 |
| 81 | 9085 | 9090 | 9096 | 9101 | 9106 | 9112 | 9117 | 9122 | 9128 | 9133 | 1 | 1 | 2 | 2 | 3 | 3 | 4 | 4 | 5 |
| 82 | 9138 | 9143 | 9149 | 9154 | 9159 | 9165 | 9170 | 9175 | 9180 | 9186 | 1 | 1 | 2 | 2 | 3 | 3 | 4 | 4 | 5 |
| 83 | 9191 | 9196 | 9201 | 9206 | 9212 | 9217 | 9222 | 9227 | 9232 | 9238 | 1 | 1 | 2 | 2 | 3 | 3 | 4 | 4 | 5 |
| 84 | 9243 | 9248 | 9253 | 9258 | 9263 | 9269 | 9274 | 9279 | 9284 | 9289 | 1 | 1 | 2 | 2 | 3 | 3 | 4 | 4 | 5 |
| 85 | 9294 | 9299 | 9304 | 9309 | 9315 | 9320 | 9325 | 9330 | 9335 | 9340 | 1 | 1 | 2 | 2 | 3 | 3 | 4 | 4 | 5 |
| 86 | 9345 | 9350 | 9355 | 9360 | 9365 | 9370 | 9375 | 9380 | 9385 | 9390 | 1 | 1 | 2 | 2 | 3 | 3 | 4 | 4 | 5 |
| 87 | 9395 | 9400 | 9405 | 9410 | 9415 | 9420 | 9425 | 9430 | 9435 | 9440 | 0 | 1 | 1 | 2 | 2 | 3 | 3 | 4 | 4 |
| 88 | 9445 | 9450 | 9455 | 9460 | 9465 | 9469 | 9474 | 9479 | 9484 | 9489 | 0 | 1 | 1 | 2 | 2 | 3 | 3 | 4 | 4 |
| 89 | 9494 | 9499 | 9504 | 9509 | 9513 | 9518 | 9523 | 9528 | 9533 | 9538 | 0 | 1 | 1 | 2 | 2 | 3 | 3 | 4 | 4 |
| 90 | 9542 | 9547 | 9552 | 9557 | 9562 | 9566 | 9571 | 9576 | 9581 | 9586 | 0 | 1 | 1 | 2 | 2 | 3 | 3 | 4 | 4 |
| 91 | 9590 | 9595 | 9600 | 9605 | 9609 | 9614 | 9619 | 9624 | 9628 | 9633 | 0 | 1 | 1 | 2 | 2 | 3 | 3 | 4 | 4 |
| 92 | 9638 | 9643 | 9647 | 9652 | 9657 | 9661 | 9666 | 9671 | 9675 | 9680 | 0 | 1 | 1 | 2 | 2 | 3 | 3 | 4 | 4 |
| 93 | 9685 | 9689 | 9694 | 9699 | 9703 | 9708 | 9713 | 9717 | 9722 | 9727 | 0 | 1 | 1 | 2 | 2 | 3 | 3 | 4 | 4 |
| 94 | 9731 | 9736 | 9741 | 9745 | 9750 | 9754 | 9759 | 9763 | 9768 | 9773 | 0 | 1 | 1 | 2 | 2 | 3 | 3 | 4 | 4 |
| 95 | 9777 | 9782 | 9786 | 9791 | 9795 | 9800 | 9805 | 9809 | 9814 | 9818 | 0 | 1 | 1 | 2 | 2 | 3 | 3 | 4 | 4 |
| 96 | 9823 | 9827 | 9832 | 9836 | 9841 | 9845 | 9850 | 9854 | 9859 | 9863 | 0 | 1 | 1 | 2 | 2 | 3 | 3 | 4 | 4 |
| 97 | 9868 | 9872 | 9877 | 9881 | 9886 | 9890 | 9894 | 9899 | 9903 | 9908 | 0 | 1 | 1 | 2 | 2 | 3 | 3 | 4 | 4 |
| 98 | 9912 | 9917 | 9921 | 9926 | 9930 | 9934 | 9939 | 9943 | 9948 | 9952 | 0 | 1 | 1 | 2 | 2 | 3 | 3 | 4 | 4 |
| 99 | 9956 | 9961 | 9965 | 9969 | 9974 | 9978 | 9983 | 9987 | 9991 | 9996 | 0 | 1 | 1 | 2 | 2 | 3 | 3 | 3 | 4 |

# SINES

| | 0′ | 10′ | 20′ | 30′ | 40′ | 50′ | 60′ | | 1′ | 2′ | 3′ | 4′ | 5′ | 6′ | 7′ | 8′ | 9′ |
|---|---|---|---|---|---|---|---|---|---|---|---|---|---|---|---|---|---|
| 0° | 0·0000 | 0·0029 | 0·0058 | 0·0087 | 0·0116 | 0·0145 | 0·0175 | 89° | 3 | 6 | 9 | 12 | 15 | 17 | 20 | 23 | 26 |
| 1 | ·0175 | ·0204 | ·0233 | ·0262 | ·0291 | ·0320 | ·0349 | 88 | 3 | 6 | 9 | 12 | 15 | 17 | 20 | 23 | 26 |
| 2 | ·0349 | ·0378 | ·0407 | ·0436 | ·0465 | ·0494 | ·0523 | 87 | 3 | 6 | 9 | 12 | 15 | 17 | 20 | 23 | 26 |
| 3 | ·0523 | ·0552 | ·0581 | ·0610 | ·0640 | ·0669 | ·0698 | 86 | 3 | 6 | 9 | 12 | 15 | 17 | 20 | 23 | 26 |
| 4 | ·0698 | ·0727 | ·0756 | ·0785 | ·0814 | ·0843 | ·0872 | 85 | 3 | 6 | 9 | 12 | 15 | 17 | 20 | 23 | 26 |
| 5 | 0·0872 | 0·0901 | 0·0929 | 0·0958 | 0·0987 | 0·1016 | 0·1045 | 84 | 3 | 6 | 9 | 12 | 14 | 17 | 20 | 23 | 26 |
| 6 | ·1045 | ·1074 | ·1103 | ·1132 | ·1161 | ·1190 | ·1219 | 83 | 3 | 6 | 9 | 12 | 14 | 17 | 20 | 23 | 26 |
| 7 | ·1219 | ·1248 | ·1276 | ·1305 | ·1334 | ·1363 | ·1392 | 82 | 3 | 6 | 9 | 12 | 14 | 17 | 20 | 23 | 26 |
| 8 | ·1392 | ·1421 | ·1449 | ·1478 | ·1507 | ·1536 | ·1564 | 81 | 3 | 6 | 9 | 11 | 14 | 17 | 20 | 23 | 26 |
| 9 | ·1564 | ·1593 | ·1622 | ·1650 | ·1679 | ·1708 | ·1736 | 80° | 3 | 6 | 9 | 11 | 14 | 17 | 20 | 23 | 26 |
| 10° | 0·1736 | 0·1765 | 0·1794 | 0·1822 | 0·1851 | 0·1880 | 0·1908 | 79 | 3 | 6 | 9 | 11 | 14 | 17 | 20 | 23 | 26 |
| 11 | ·1908 | ·1937 | ·1965 | ·1994 | ·2022 | ·2051 | ·2079 | 78 | 3 | 6 | 9 | 11 | 14 | 17 | 20 | 23 | 26 |
| 12 | ·2079 | ·2108 | ·2136 | ·2164 | ·2193 | ·2221 | ·2250 | 77 | 3 | 6 | 9 | 11 | 14 | 17 | 20 | 23 | 26 |
| 13 | ·2250 | ·2278 | ·2306 | ·2334 | ·2363 | ·2391 | ·2419 | 76 | 3 | 6 | 8 | 11 | 14 | 17 | 20 | 23 | 25 |
| 14 | ·2419 | ·2447 | ·2476 | ·2504 | ·2532 | ·2560 | ·2588 | 75 | 3 | 6 | 8 | 11 | 14 | 17 | 20 | 23 | 25 |
| 15 | 0·2588 | 0·2616 | 0·2644 | 0·2672 | 0·2700 | 0·2728 | 0·2756 | 74 | 3 | 6 | 8 | 11 | 14 | 17 | 20 | 22 | 25 |
| 16 | ·2756 | ·2784 | ·2812 | ·2840 | ·2868 | ·2896 | ·2924 | 73 | 3 | 6 | 8 | 11 | 14 | 17 | 20 | 22 | 25 |
| 17 | ·2924 | ·2952 | ·2979 | ·3007 | ·3035 | ·3062 | ·3090 | 72 | 3 | 6 | 8 | 11 | 14 | 17 | 19 | 22 | 25 |
| 18 | ·3090 | ·3118 | ·3145 | ·3173 | ·3201 | ·3228 | ·3256 | 71 | 3 | 6 | 8 | 11 | 14 | 17 | 19 | 22 | 25 |
| 19 | ·3256 | ·3283 | ·3311 | ·3338 | ·3365 | ·3393 | ·3420 | 70° | 3 | 5 | 8 | 11 | 14 | 16 | 19 | 22 | 25 |
| 20° | 0·3420 | 0·3448 | 0·3475 | 0·3502 | 0·3529 | 0·3557 | 0·3584 | 69 | 3 | 5 | 8 | 11 | 14 | 16 | 19 | 22 | 25 |
| 21 | ·3584 | ·3611 | ·3638 | ·3665 | ·3692 | ·3719 | ·3746 | 68 | 3 | 5 | 8 | 11 | 14 | 16 | 19 | 22 | 24 |
| 22 | ·3746 | ·3773 | ·3800 | ·3827 | ·3854 | ·3881 | ·3907 | 67 | 3 | 5 | 8 | 11 | 13 | 16 | 19 | 21 | 24 |
| 23 | ·3907 | ·3934 | ·3961 | ·3987 | ·4014 | ·4041 | ·4067 | 66 | 3 | 5 | 8 | 11 | 13 | 16 | 19 | 21 | 24 |
| 24 | ·4067 | ·4094 | ·4120 | ·4147 | ·4173 | ·4200 | ·4226 | 65 | 3 | 5 | 8 | 11 | 13 | 16 | 19 | 21 | 24 |
| 25 | 0·4226 | 0·4253 | 0·4279 | 0·4305 | 0·4331 | 0·4358 | 0·4384 | 64 | 3 | 5 | 8 | 10 | 13 | 16 | 18 | 21 | 24 |
| 26 | ·4384 | ·4410 | ·4436 | ·4462 | ·4488 | ·4514 | ·4540 | 63 | 3 | 5 | 8 | 10 | 13 | 16 | 18 | 21 | 23 |
| 27 | ·4540 | ·4566 | ·4592 | ·4617 | ·4643 | ·4669 | ·4695 | 62 | 3 | 5 | 8 | 10 | 13 | 15 | 18 | 21 | 23 |
| 28 | ·4695 | ·4720 | ·4746 | ·4772 | ·4797 | ·4823 | ·4848 | 61 | 3 | 5 | 8 | 10 | 13 | 15 | 18 | 20 | 23 |
| 29 | ·4848 | ·4874 | ·4899 | ·4924 | ·4950 | ·4975 | ·5000 | 60° | 3 | 5 | 8 | 10 | 13 | 15 | 18 | 20 | 23 |
| 30° | 0·5000 | 0·5025 | 0·5050 | 0·5075 | 0·5100 | 0·5125 | 0·5150 | 59 | 3 | 5 | 8 | 10 | 13 | 15 | 18 | 20 | 23 |
| 31 | ·5150 | ·5175 | ·5200 | ·5225 | ·5250 | ·5275 | ·5299 | 58 | 2 | 5 | 7 | 10 | 12 | 15 | 17 | 20 | 22 |
| 32 | ·5299 | ·5324 | ·5348 | ·5373 | ·5398 | ·5422 | ·5446 | 57 | 2 | 5 | 7 | 10 | 12 | 15 | 17 | 20 | 22 |
| 33 | ·5446 | ·5471 | ·5495 | ·5519 | ·5544 | ·5568 | ·5592 | 56 | 2 | 5 | 7 | 10 | 12 | 15 | 17 | 19 | 22 |
| 34 | ·5592 | ·5616 | ·5640 | ·5664 | ·5688 | ·5712 | ·5736 | 55 | 2 | 5 | 7 | 10 | 12 | 14 | 17 | 19 | 22 |
| 35 | 0·5736 | 0·5760 | 0·5783 | 0·5807 | 0·5831 | 0·5854 | 0·5878 | 54 | 2 | 5 | 7 | 9 | 12 | 14 | 17 | 19 | 21 |
| 36 | ·5878 | ·5901 | ·5925 | ·5948 | ·5972 | ·5995 | ·6018 | 53 | 2 | 5 | 7 | 9 | 12 | 14 | 16 | 19 | 21 |
| 37 | ·6018 | ·6041 | ·6065 | ·6088 | ·6111 | ·6134 | ·6157 | 52 | 2 | 5 | 7 | 9 | 12 | 14 | 16 | 18 | 21 |
| 38 | ·6157 | ·6180 | ·6202 | ·6225 | ·6248 | ·6271 | ·6293 | 51 | 2 | 5 | 7 | 9 | 11 | 14 | 16 | 18 | 20 |
| 39 | ·6293 | ·6316 | ·6338 | ·6361 | ·6383 | ·6406 | ·6428 | 50° | 2 | 4 | 7 | 9 | 11 | 13 | 16 | 18 | 20 |
| 40° | 0·6428 | 0·6450 | 0·6472 | 0·6494 | 0·6517 | 0·6539 | 0·6561 | 49 | 2 | 4 | 7 | 9 | 11 | 13 | 15 | 18 | 20 |
| 41 | ·6561 | ·6583 | ·6604 | ·6626 | ·6648 | ·6670 | ·6691 | 48 | 2 | 4 | 7 | 9 | 11 | 13 | 15 | 17 | 20 |
| 42 | ·6691 | ·6713 | ·6734 | ·6756 | ·6777 | ·6799 | ·6820 | 47 | 2 | 4 | 6 | 9 | 11 | 13 | 15 | 17 | 19 |
| 43 | ·6820 | ·6841 | ·6862 | ·6884 | ·6905 | ·6926 | ·6947 | 46 | 2 | 4 | 6 | 8 | 11 | 13 | 15 | 17 | 19 |
| 44 | ·6947 | ·6967 | ·6988 | ·7009 | ·7030 | ·7050 | ·7071 | 45 | 2 | 4 | 6 | 8 | 10 | 12 | 15 | 17 | 19 |
| | 60′ | 50′ | 40′ | 30′ | 20′ | 10′ | 0′ | | 1′ | 2′ | 3′ | 4′ | 5′ | 6′ | 7′ | 8′ | 9′ |

Differences to be subtracted

# COSINES

# SINES

| | 0′ | 10′ | 20′ | 30′ | 40′ | 50′ | 60′ | | 1′ | 2′ | 3′ | 4′ | 5′ | 6′ | 7′ | 8′ | 9′ |
|---|---|---|---|---|---|---|---|---|---|---|---|---|---|---|---|---|---|
| 45° | 0·7071 | 0·7092 | 0·7112 | 0·7133 | 0·7153 | 0·7173 | 0·7193 | 44° | 2 | 4 | 6 | 8 | 10 | 12 | 14 | 16 | 18 |
| 46 | ·7193 | ·7214 | ·7234 | ·7254 | ·7274 | ·7294 | ·7314 | 43 | 2 | 4 | 6 | 8 | 10 | 12 | 14 | 16 | 18 |
| 47 | ·7314 | ·7333 | ·7353 | ·7373 | ·7392 | ·7412 | ·7431 | 42 | 2 | 4 | 6 | 8 | 10 | 12 | 14 | 16 | 18 |
| 48 | ·7431 | ·7451 | ·7470 | ·7490 | ·7509 | ·7528 | ·7547 | 41 | 2 | 4 | 6 | 8 | 10 | 12 | 13 | 15 | 17 |
| 49 | ·7547 | ·7566 | ·7585 | ·7604 | ·7623 | ·7642 | ·7660 | 40° | 2 | 4 | 6 | 8 | 9 | 11 | 13 | 15 | 17 |
| 50° | 0·7660 | 0·7679 | 0·7698 | 0·7716 | 0·7735 | 0·7753 | 0·7771 | 39 | 2 | 4 | 6 | 7 | 9 | 11 | 13 | 15 | 17 |
| 51 | ·7771 | ·7790 | ·7808 | ·7826 | ·7844 | ·7862 | ·7880 | 38 | 2 | 4 | 5 | 7 | 9 | 11 | 13 | 14 | 16 |
| 52 | ·7880 | ·7898 | ·7916 | ·7934 | ·7951 | ·7969 | ·7986 | 37 | 2 | 4 | 5 | 7 | 9 | 11 | 12 | 14 | 16 |
| 53 | ·7986 | ·8004 | ·8021 | ·8039 | ·8056 | ·8073 | ·8090 | 36 | 2 | 3 | 5 | 7 | 9 | 10 | 12 | 14 | 16 |
| 54 | ·8090 | ·8107 | ·8124 | ·8141 | ·8158 | ·8175 | ·8192 | 35 | 2 | 3 | 5 | 7 | 8 | 10 | 12 | 14 | 15 |
| 55 | 0·8192 | 0·8208 | 0·8225 | 0·8241 | 0·8258 | 0·8274 | 0·8290 | 34 | 2 | 3 | 5 | 7 | 8 | 10 | 12 | 13 | 15 |
| 56 | ·8290 | ·8307 | ·8323 | ·8339 | ·8355 | ·8371 | ·8387 | 33 | 2 | 3 | 5 | 6 | 8 | 10 | 11 | 13 | 14 |
| 57 | ·8387 | ·8403 | ·8418 | ·8434 | ·8450 | ·8465 | ·8480 | 32 | 2 | 3 | 5 | 6 | 8 | 9 | 11 | 13 | 14 |
| 58 | ·8480 | ·8496 | ·8511 | ·8526 | ·8542 | ·8557 | ·8572 | 31 | 2 | 3 | 5 | 6 | 8 | 9 | 11 | 12 | 14 |
| 59 | ·8572 | ·8587 | ·8601 | ·8616 | ·8631 | ·8646 | ·8660 | 30° | 1 | 3 | 4 | 6 | 7 | 9 | 10 | 12 | 13 |
| 60° | 0·8660 | 0·8675 | 0·8689 | 0·8704 | 0·8718 | 0·8732 | 0·8746 | 29 | 1 | 3 | 4 | 6 | 7 | 9 | 10 | 11 | 13 |
| 61 | ·8746 | ·8760 | ·8774 | ·8788 | ·8802 | ·8816 | ·8829 | 28 | 1 | 3 | 4 | 6 | 7 | 8 | 10 | 11 | 12 |
| 62 | ·8829 | ·8843 | ·8857 | ·8870 | ·8884 | ·8897 | ·8910 | 27 | 1 | 3 | 4 | 5 | 7 | 8 | 9 | 11 | 12 |
| 63 | ·8910 | ·8923 | ·8936 | ·8949 | ·8962 | ·8975 | ·8988 | 26 | 1 | 3 | 4 | 5 | 6 | 8 | 9 | 10 | 12 |
| 64 | ·8988 | ·9001 | ·9013 | ·9026 | ·9038 | ·9051 | ·9063 | 25 | 1 | 3 | 4 | 5 | 6 | 8 | 9 | 10 | 11 |
| 65 | 0·9063 | 0·9075 | 0·9088 | 0·9100 | 0·9112 | 0·9124 | 0·9135 | 24 | 1 | 2 | 4 | 5 | 6 | 7 | 8 | 10 | 11 |
| 66 | ·9135 | ·9147 | ·9159 | ·9171 | ·9182 | ·9194 | ·9205 | 23 | 1 | 2 | 3 | 5 | 6 | 7 | 8 | 9 | 10 |
| 67 | ·9205 | ·9216 | ·9228 | ·9239 | ·9250 | ·9261 | ·9272 | 22 | 1 | 2 | 3 | 4 | 6 | 7 | 8 | 9 | 10 |
| 68 | ·9272 | ·9283 | ·9293 | ·9304 | ·9315 | ·9325 | ·9336 | 21 | 1 | 2 | 3 | 4 | 5 | 6 | 7 | 9 | 10 |
| 69 | ·9336 | ·9346 | ·9356 | ·9367 | ·9377 | ·9387 | ·9397 | 20° | 1 | 2 | 3 | 4 | 5 | 6 | 7 | 8 | 9 |
| 70° | 0·9397 | 0·9407 | 0·9417 | 0·9426 | 0·9436 | 0·9446 | 0·9455 | 19 | 1 | 2 | 3 | 4 | 5 | 6 | 7 | 8 | 9 |
| 71 | ·9455 | ·9465 | ·9474 | ·9483 | ·9492 | ·9502 | ·9511 | 18 | 1 | 2 | 3 | 4 | 5 | 6 | 6 | 7 | 8 |
| 72 | ·9511 | ·9520 | ·9528 | ·9537 | ·9546 | ·9555 | ·9563 | 17 | 1 | 2 | 3 | 4 | 4 | 5 | 6 | 7 | 8 |
| 73 | ·9563 | ·9572 | ·9580 | ·9588 | ·9596 | ·9605 | ·9613 | 16 | 1 | 2 | 2 | 3 | 4 | 5 | 6 | 7 | 7 |
| 74 | ·9613 | ·9621 | ·9628 | ·9636 | ·9644 | ·9652 | ·9659 | 15 | 1 | 2 | 2 | 3 | 4 | 5 | 5 | 6 | 7 |
| 75 | 0·9659 | 0·9667 | 0·9674 | 0·9681 | 0·9689 | 0·9696 | 0·9703 | 14 | 1 | 1 | 2 | 3 | 4 | 4 | 5 | 6 | 7 |
| 76 | ·9703 | ·9710 | ·9717 | ·9724 | ·9730 | ·9737 | ·9744 | 13 | 1 | 1 | 2 | 3 | 3 | 4 | 5 | 5 | 6 |
| 77 | ·9744 | ·9750 | ·9757 | ·9763 | ·9769 | ·9775 | ·9781 | 12 | 1 | 1 | 2 | 3 | 3 | 4 | 4 | 5 | 6 |
| 78 | ·9781 | ·9787 | ·9793 | ·9799 | ·9805 | ·9811 | ·9816 | 11 | 1 | 1 | 2 | 2 | 3 | 3 | 4 | 5 | 5 |
| 79 | ·9816 | ·9822 | ·9827 | ·9833 | ·9838 | ·9843 | ·9848 | 10° | 1 | 1 | 2 | 2 | 3 | 3 | 4 | 4 | 5 |
| 80° | 0·9848 | 0·9853 | 0·9858 | 0·9863 | 0·9868 | 0·9872 | 0·9877 | 9 | 0 | 1 | 1 | 2 | 2 | 3 | 3 | 4 | 4 |
| 81 | ·9877 | ·9881 | ·9886 | ·9890 | ·9894 | ·9899 | ·9903 | 8 | 0 | 1 | 1 | 2 | 2 | 3 | 3 | 4 | 4 |
| 82 | ·9903 | ·9907 | ·9911 | ·9914 | ·9918 | ·9922 | ·9925 | 7 | 0 | 1 | 1 | 2 | 2 | 2 | 3 | 3 | 3 |
| 83 | ·9925 | ·9929 | ·9932 | ·9936 | ·9939 | ·9942 | ·9945 | 6 | 0 | 1 | 1 | 1 | 2 | 2 | 2 | 3 | 3 |
| 84 | ·9945 | ·9948 | ·9951 | ·9954 | ·9957 | ·9959 | ·9962 | 5 | 0 | 1 | 1 | 1 | 1 | 2 | 2 | 2 | 3 |
| 85 | 0·9962 | 0·9964 | 0·9967 | 0·9969 | 0·9971 | 0·9974 | 0·9976 | 4 | 0 | 0 | 1 | 1 | 1 | 1 | 2 | 2 | 2 |
| 86 | ·9976 | ·9978 | ·9980 | ·9981 | ·9983 | ·9985 | ·9986 | 3 | 0 | 0 | 0 | 1 | 1 | 1 | 1 | 1 | 2 |
| 87 | ·9986 | ·9988 | ·9989 | ·9990 | ·9992 | ·9993 | ·9994 | 2 | 0 | 0 | 0 | 0 | 0 | 1 | 1 | 1 | 1 |
| 88 | ·9994 | ·9995 | ·9996 | ·9997 | ·9997 | ·9998 | ·9998 | 1 | 0 | 0 | 0 | 0 | 0 | 0 | 0 | 0 | 0 |
| 89 | 0·9998 | 0·9999 | 0·9999 | 1·0000 | 1·0000 | 1·0000 | 1·0000 | 0° | 0 | 0 | 0 | 0 | 0 | 0 | 0 | 0 | 0 |
| | 60′ | 50′ | 40′ | 30′ | 20′ | 10′ | 0′ | | 1′ | 2′ | 3′ | 4′ | 5′ | 6′ | 7′ | 8′ | 9′ |

Differences to be subtracted

# COSINES

[ 151 ]

# TANGENTS

| | 0' | 10' | 20' | 30' | 40' | 50' | 60' | | 1' | 2' | 3' | 4' | 5' | 6' | 7' | 8' | 9' |
|---|---|---|---|---|---|---|---|---|---|---|---|---|---|---|---|---|---|
| 0° | 0·0000 | 0·0029 | 0·0058 | 0·0087 | 0·0116 | 0·0145 | 0·0175 | 89° | 3 | 6 | 9 | 12 | 15 | 17 | 20 | 23 | 26 |
| 1 | ·0175 | ·0204 | ·0233 | ·0262 | ·0291 | ·0320 | ·0349 | 88 | 3 | 6 | 9 | 12 | 15 | 17 | 20 | 23 | 26 |
| 2 | ·0349 | ·0378 | ·0407 | ·0437 | ·0466 | ·0495 | ·0524 | 87 | 3 | 6 | 9 | 12 | 15 | 18 | 20 | 23 | 26 |
| 3 | ·0524 | ·0553 | ·0582 | ·0612 | ·0641 | ·0670 | ·0699 | 86 | 3 | 6 | 9 | 12 | 15 | 18 | 20 | 23 | 26 |
| 4 | ·0699 | ·0729 | ·0758 | ·0787 | ·0816 | ·0846 | ·0875 | 85 | 3 | 6 | 9 | 12 | 15 | 18 | 21 | 23 | 26 |
| 5 | 0·0875 | 0·0904 | 0·0934 | 0·0963 | 0·0992 | 0·1022 | 0·1051 | 84 | 3 | 6 | 9 | 12 | 15 | 18 | 21 | 24 | 26 |
| 6 | ·1051 | ·1080 | ·1110 | ·1139 | ·1169 | ·1198 | ·1228 | 83 | 3 | 6 | 9 | 12 | 15 | 18 | 21 | 24 | 27 |
| 7 | ·1228 | ·1257 | ·1287 | ·1317 | ·1346 | ·1376 | ·1405 | 82 | 3 | 6 | 9 | 12 | 15 | 18 | 21 | 24 | 27 |
| 8 | ·1405 | ·1435 | ·1465 | ·1495 | ·1524 | ·1554 | ·1584 | 81 | 3 | 6 | 9 | 12 | 15 | 18 | 21 | 24 | 27 |
| 9 | ·1584 | ·1614 | ·1644 | ·1673 | ·1703 | ·1733 | ·1763 | 80° | 3 | 6 | 9 | 12 | 15 | 18 | 21 | 24 | 27 |
| 10° | 0·1763 | 0·1793 | 0·1823 | 0·1853 | 0·1883 | 0·1914 | 0·1944 | 79 | 3 | 6 | 9 | 12 | 15 | 18 | 21 | 24 | 27 |
| 11 | ·1944 | ·1974 | ·2004 | ·2035 | ·2065 | ·2095 | ·2126 | 78 | 3 | 6 | 9 | 12 | 15 | 18 | 21 | 24 | 27 |
| 12 | ·2126 | ·2156 | ·2186 | ·2217 | ·2247 | ·2278 | ·2309 | 77 | 3 | 6 | 9 | 12 | 15 | 18 | 21 | 24 | 27 |
| 13 | ·2309 | ·2339 | ·2370 | ·2401 | ·2432 | ·2462 | ·2493 | 76 | 3 | 6 | 9 | 12 | 15 | 18 | 22 | 25 | 28 |
| 14 | ·2493 | ·2524 | ·2555 | ·2586 | ·2617 | ·2648 | ·2679 | 75 | 3 | 6 | 9 | 12 | 16 | 19 | 22 | 25 | 28 |
| 15 | 0·2679 | 0·2711 | 0·2742 | 0·2773 | 0·2805 | 0·2836 | 0·2867 | 74 | 3 | 6 | 9 | 13 | 16 | 19 | 22 | 25 | 28 |
| 16 | ·2867 | ·2899 | ·2931 | ·2962 | ·2994 | ·3026 | ·3057 | 73 | 3 | 6 | 9 | 13 | 16 | 19 | 22 | 25 | 28 |
| 17 | ·3057 | ·3089 | ·3121 | ·3153 | ·3185 | ·3217 | ·3249 | 72 | 3 | 6 | 10 | 13 | 16 | 19 | 22 | 26 | 29 |
| 18 | ·3249 | ·3281 | ·3314 | ·3346 | ·3378 | ·3411 | ·3443 | 71 | 3 | 6 | 10 | 13 | 16 | 19 | 23 | 26 | 29 |
| 19 | ·3443 | ·3476 | ·3508 | ·3541 | ·3574 | ·3607 | ·3640 | 70° | 3 | 7 | 10 | 13 | 16 | 20 | 23 | 26 | 29 |
| 20° | 0·3640 | 0·3673 | 0·3706 | 0·3739 | 0·3772 | 0·3805 | 0·3839 | 69 | 3 | 7 | 10 | 13 | 17 | 20 | 23 | 27 | 30 |
| 21 | ·3839 | ·3872 | ·3906 | ·3939 | ·3973 | ·4006 | ·4040 | 68 | 3 | 7 | 10 | 13 | 17 | 20 | 24 | 27 | 30 |
| 22 | ·4040 | ·4074 | ·4108 | ·4142 | ·4176 | ·4210 | ·4245 | 67 | 3 | 7 | 10 | 14 | 17 | 20 | 24 | 27 | 31 |
| 23 | ·4245 | ·4279 | ·4314 | ·4348 | ·4383 | ·4417 | ·4452 | 66 | 3 | 7 | 10 | 14 | 17 | 21 | 24 | 28 | 31 |
| 24 | ·4452 | ·4487 | ·4522 | ·4557 | ·4592 | ·4628 | ·4663 | 65 | 4 | 7 | 11 | 14 | 18 | 21 | 25 | 28 | 32 |
| 25 | 0·4663 | 0·4699 | 0·4734 | 0·4770 | 0·4806 | 0·4841 | 0·4877 | 64 | 4 | 7 | 11 | 14 | 18 | 21 | 25 | 29 | 32 |
| 26 | ·4877 | ·4913 | ·4950 | ·4986 | ·5022 | ·5059 | ·5095 | 63 | 4 | 7 | 11 | 15 | 18 | 22 | 25 | 29 | 33 |
| 27 | ·5095 | ·5132 | ·5169 | ·5206 | ·5243 | ·5280 | ·5317 | 62 | 4 | 7 | 11 | 15 | 18 | 22 | 26 | 30 | 33 |
| 28 | ·5317 | ·5354 | ·5392 | ·5430 | ·5467 | ·5505 | ·5543 | 61 | 4 | 8 | 11 | 15 | 19 | 23 | 26 | 30 | 34 |
| 29 | ·5543 | ·5581 | ·5619 | ·5658 | ·5696 | ·5735 | ·5774 | 60° | 4 | 8 | 12 | 15 | 19 | 23 | 27 | 31 | 35 |
| 30° | 0·5774 | 0·5812 | 0·5851 | 0·5890 | 0·5930 | 0·5969 | 0·6009 | 59 | 4 | 8 | 12 | 16 | 20 | 24 | 27 | 31 | 35 |
| 31 | ·6009 | ·6048 | ·6088 | ·6128 | ·6168 | ·6208 | ·6249 | 58 | 4 | 8 | 12 | 16 | 20 | 24 | 28 | 32 | 36 |
| 32 | ·6249 | ·6289 | ·6330 | ·6371 | ·6412 | ·6453 | ·6494 | 57 | 4 | 8 | 12 | 16 | 20 | 25 | 29 | 33 | 37 |
| 33 | ·6494 | ·6536 | ·6577 | ·6619 | ·6661 | ·6703 | ·6745 | 56 | 4 | 8 | 13 | 17 | 21 | 25 | 29 | 33 | 38 |
| 34 | ·6745 | ·6787 | ·6830 | ·6873 | ·6916 | ·6959 | ·7002 | 55 | 4 | 9 | 13 | 17 | 21 | 26 | 30 | 34 | 39 |
| 35 | 0·7002 | 0·7046 | 0·7089 | 0·7133 | 0·7177 | 0·7221 | 0·7265 | 54 | 4 | 9 | 13 | 18 | 22 | 26 | 31 | 35 | 40 |
| 36 | ·7265 | ·7310 | ·7355 | ·7400 | ·7445 | ·7490 | ·7536 | 53 | 5 | 9 | 14 | 18 | 23 | 27 | 32 | 36 | 41 |
| 37 | ·7536 | ·7581 | ·7627 | ·7673 | ·7720 | ·7766 | ·7813 | 52 | 5 | 9 | 14 | 18 | 23 | 28 | 32 | 37 | 42 |
| 38 | ·7813 | ·7860 | ·7907 | ·7954 | ·8002 | ·8050 | ·8098 | 51 | 5 | 10 | 14 | 19 | 24 | 29 | 33 | 38 | 43 |
| 39 | ·8098 | ·8146 | ·8195 | ·8243 | ·8292 | ·8342 | ·8391 | 50° | 5 | 10 | 15 | 20 | 24 | 29 | 34 | 39 | 44 |
| 40° | 0·8391 | 0·8441 | 0·8491 | 0·8541 | 0·8591 | 0·8642 | 0·8693 | 49 | 5 | 10 | 15 | 20 | 25 | 30 | 35 | 40 | 45 |
| 41 | ·8603 | ·8744 | ·8796 | ·8847 | ·8899 | ·8952 | ·9004 | 48 | 5 | 10 | 16 | 21 | 26 | 31 | 36 | 41 | 47 |
| 42 | ·9004 | ·9057 | ·9110 | ·9163 | ·9217 | ·9271 | ·9325 | 47 | 5 | 11 | 16 | 21 | 27 | 32 | 37 | 43 | 48 |
| 43 | ·9325 | ·9380 | ·9435 | ·9490 | ·9545 | ·9601 | ·9657 | 46 | 6 | 11 | 17 | 22 | 28 | 33 | 39 | 44 | 50 |
| 44 | ·9657 | ·9713 | ·9770 | ·9827 | ·9884 | ·9942 | 1·0000 | 45 | 6 | 11 | 17 | 23 | 29 | 34 | 40 | 46 | 51 |
| | 60' | 50' | 40' | 30' | 20' | 10' | 0' | | 1' | 2' | 3' | 4' | 5' | 6' | 7' | 8' | 9' |

Differences to be subtracted

# COTANGENTS

# TANGENTS

| | 0′ | 10′ | 20′ | 30′ | 40′ | 50′ | 60′ | | 1′ | 2′ | 3′ | 4′ | 5′ | 6′ | 7′ | 8′ | 9′ |
|---|---|---|---|---|---|---|---|---|---|---|---|---|---|---|---|---|---|
| 45° | 1·0000 | 1·0058 | 1·0117 | 1·0176 | 1·0235 | 1·0295 | 1·0355 | 44° | 6 | 12 | 18 | 24 | 30 | 36 | 41 | 47 | 53 |
| 46 | ·0355 | ·0416 | ·0477 | ·0538 | ·0599 | ·0661 | ·0724 | 43 | 6 | 12 | 18 | 25 | 31 | 37 | 43 | 49 | 55 |
| 47 | ·0724 | ·0786 | ·0850 | ·0913 | ·0977 | ·1041 | ·1106 | 42 | 6 | 13 | 19 | 26 | 32 | 38 | 45 | 51 | 57 |
| 48 | ·1106 | ·1171 | ·1237 | ·1303 | ·1369 | ·1436 | ·1504 | 41 | 7 | 13 | 20 | 27 | 33 | 40 | 46 | 53 | 60 |
| 49 | ·1504 | ·1571 | ·1640 | ·1708 | ·1778 | ·1847 | ·1918 | 40° | 7 | 14 | 21 | 28 | 34 | 41 | 48 | 55 | 62 |
| 50° | 1·1918 | 1·1988 | 1·2059 | 1·2131 | 1·2203 | 1·2276 | 1·2349 | 39 | 7 | 14 | 22 | 29 | 36 | 43 | 50 | 58 | 65 |
| 51 | ·2349 | ·2423 | ·2497 | ·2572 | ·2647 | ·2723 | ·2799 | 38 | 8 | 15 | 23 | 30 | 38 | 45 | 53 | 60 | 68 |
| 52 | ·2799 | ·2876 | ·2954 | ·3032 | ·3111 | ·3190 | ·3270 | 37 | 8 | 16 | 24 | 31 | 39 | 47 | 55 | 63 | 71 |
| 53 | ·3270 | ·3351 | ·3432 | ·3514 | ·3597 | ·3680 | ·3764 | 36 | 8 | 16 | 25 | 33 | 41 | 49 | 58 | 66 | 74 |
| 54 | ·3764 | ·3848 | ·3934 | ·4019 | ·4106 | ·4193 | ·4281 | 35 | 9 | 17 | 26 | 35 | 43 | 52 | 60 | 69 | 78 |
| 55 | 1·4281 | 1·4370 | 1·4460 | 1·4550 | 1·4641 | 1·4733 | 1·4826 | 34 | 9 | 18 | 27 | 36 | 45 | 54 | 63 | 73 | 82 |
| 56 | ·4826 | ·4919 | ·5013 | ·5108 | ·5204 | ·5301 | ·5399 | 33 | 10 | 19 | 29 | 38 | 48 | 57 | 67 | 76 | 86 |
| 57 | ·5399 | ·5497 | ·5597 | ·5697 | ·5798 | ·5900 | ·6003 | 32 | 10 | 20 | 30 | 40 | 50 | 60 | 71 | 81 | 91 |
| 58 | ·6003 | ·6107 | ·6212 | ·6319 | ·6426 | ·6534 | ·6643 | 31 | 11 | 21 | 32 | 43 | 53 | 64 | 75 | 85 | 96 |
| 59 | ·6643 | ·6753 | ·6864 | ·6977 | ·7090 | ·7205 | ·7321 | 30° | 11 | 23 | 34 | 45 | 56 | 68 | 79 | 90 | 102 |
| 60° | 1·732 | 1·744 | 1·756 | 1·767 | 1·780 | 1·792 | 1·804 | 29 | 1 | 2 | 4 | 5 | 6 | 7 | 8 | 10 | 11 |
| 61 | 1·804 | 1·816 | 1·829 | 1·842 | 1·855 | 1·868 | 1·881 | 28 | 1 | 3 | 4 | 5 | 6 | 8 | 9 | 10 | 12 |
| 62 | 1·881 | 1·894 | 1·907 | 1·921 | 1·935 | 1·949 | 1·963 | 27 | 1 | 3 | 4 | 5 | 7 | 8 | 10 | 11 | 12 |
| 63 | 1·963 | 1·977 | 1·991 | 2·006 | 2·020 | 2·035 | 2·050 | 26 | 1 | 3 | 4 | 6 | 7 | 9 | 10 | 12 | 13 |
| 64 | 2·050 | 2·066 | 2·081· | 2·097 | 2·112 | 2·128 | 2·145 | 25 | 2 | 3 | 5 | 6 | 8 | 9 | 11 | 13 | 14 |
| 65 | 2·145 | 2·161 | 2·177 | 2·194 | 2·211 | 2·229 | 2·246 | 24 | 2 | 3 | 5 | 7 | 8 | 10 | 12 | 14 | 15 |
| 66 | 2·246 | 2·264 | 2·282 | 2·300 | 2·318 | 2·337 | 2·356 | 23 | 2 | 4 | 5 | 7 | 9 | 11 | 13 | 15 | 16 |
| 67 | 2·356 | 2·375 | 2·394 | 2·414 | 2·434 | 2·455 | 2·475 | 22 | 2 | 4 | 6 | 8 | 10 | 12 | 14 | 16 | 18 |
| 68 | 2·475 | 2·496 | 2·517 | 2·539 | 2·560 | 2·583 | 2·605 | 21 | 2 | 4 | 6 | 9 | 11 | 13 | 15 | 17 | 20 |
| 69 | 2·605 | 2·628 | 2·651 | 2·675 | 2·699 | 2·723 | 2·747 | 20° | 2 | 5 | 7 | 9 | 12 | 14 | 17 | 19 | 21 |
| 70° | 2·747 | 2·773 | 2·798 | 2·824 | 2·850 | 2·877 | 2·904 | 19 | 3 | 5 | 8 | 10 | 13 | 16 | 18 | 21 | 23 |
| 71 | 2·904 | 2·932 | 2·960 | 2·989 | 3·018 | 3·047 | 3·078 | 18 | 3 | 6 | 9 | 12 | 14 | 17 | 20 | 23 | 26 |
| 72 | 3·078 | 3·108 | 3·140 | 3·172 | 3·204 | 3·237 | 3·271 | 17 | 3 | 6 | 10 | 13 | 16 | 19 | 23 | 26 | 29 |
| 73 | 3·271 | 3·305 | 3·340 | 3·376 | 3·412 | 3·450 | 3·487 | 16 | 4 | 7 | 11 | 14 | 18 | 22 | 25 | 29 | 32 |
| 74 | 3·487 | 3·526 | 3·566 | 3·606 | 3·647 | 3·689 | 3·732 | 15 | 4 | 8 | 12 | 16 | 20 | 24 | 29 | 33 | 37 |
| 75 | 3·732 | 3·776 | 3·821 | 3·867 | 3·914 | 3·962 | 4·011 | 14 | 5 | 9 | 14 | 19 | 23 | 28 | 33 | 37 | 42 |
| 76 | 4·011 | 4·061 | 4·113 | 4·165 | 4·219 | 4·275 | 4·331 | 13 | 5 | 11 | 16 | 21 | 27 | 32 | 37 | 43 | 48 |
| 77 | 4·331 | 4·390 | 4·449 | 4·511 | 4·574 | 4·638 | 4·705 | 12 | 6 | 12 | 19 | 25 | 31 | 37 | 44 | 50 | 56 |
| 78 | 4·705 | 4·773 | 4·843 | 4·915 | 4·989 | 5·066 | 5·145 | 11 | 7 | 15 | 22 | 29 | 37 | 44 | 51 | 59 | 66 |
| 79 | 5·145 | 5·226 | 5·309 | 5·396 | 5·485 | 5·576 | 5·671 | 10° | 9 | 18 | 26 | 35 | 44 | 53 | 61 | 70 | 79 |
| 80° | 5·671 | 5·769 | 5·871 | 5·976 | 6·084 | 6·197 | 6·314 | 9 | | | | | | | | | |
| 81 | 6·314 | 6·435 | 6·561 | 6·691 | 6·827 | 6·968 | 7·115 | 8 | | | | | | | | | |
| 82 | 7·115 | 7·269 | 7·429 | 7·596 | 7·770 | 7·953 | 8·144 | 7 | | | | | | | | | |
| 83 | 8·144 | 8·345 | 8·556 | 8·777 | 9·010 | 9·255 | 9·514 | 6 | | | | | | | | | |
| 84 | 9·514 | 9·788 | 10·078 | 10·385 | 10·712 | 11·059 | 11·430 | 5 | | | | | | | | | |
| 85 | 11·43 | 11·83 | 12·25 | 12·71 | 13·20 | 13·73 | 14·30 | 4 | | | | | | | | | |
| 86 | 14·30 | 14·92 | 15·60 | 16·35 | 17·17 | 18·07 | 19·08 | 3 | | | | | | | | | |
| 87 | 19·08 | 20·21 | 21·47 | 22·90 | 24·54 | 26·43 | 28·64 | 2 | | | | | | | | | |
| 88 | 28·64 | 31·24 | 34·37 | 38·19 | 42·96 | 49·10 | 57·29 | 1 | | | | | | | | | |
| 89 | 57·29 | 68·75 | 85·94 | 114·59 | 171·89 | 343·77 | ∞ | 0° | | | | | | | | | |
| | 60′ | 50′ | 40′ | 30′ | 20′ | 10′ | 0′ | | 1′ | 2′ | 3′ | 4′ | 5′ | 6′ | 7′ | 8′ | 9′ |

The differences change so rapidly here that they cannot be tabulated

Differences to be subtracted

# COTANGENTS

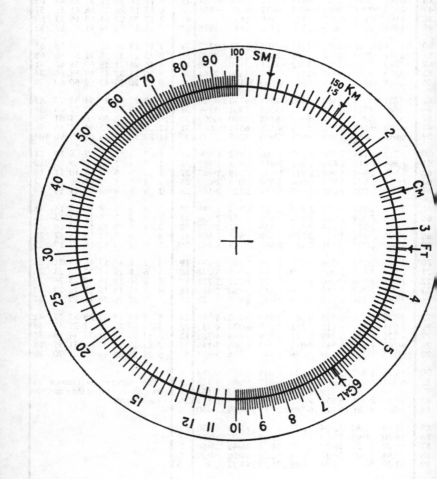

Printed in the United States
By Bookmasters